Supply Chain Management on Demand

Chae An · Hansjörg Fromm (Eds.)

Supply Chain Management on Demand

Strategies, Technologies, Applications

With contributions by numerous experts

With 87 Figures

 Springer

Chae An
IBM Software Group
294 Route 100
Somers, NY 10589
USA
chaean@us.ibm.com

Hansjörg Fromm
IBM Business Consulting Services
Pascalstr. 100
70569 Stuttgart
Germany
fromm@de.ibm.com

Library of Congress Control Number: 2005920704

ISBN-10   3-540-24423-9 Springer Berlin Heidelberg New York
ISBN-13   978-3-540-24423-3 Springer Berlin Heidelberg New York

Springer is a part of Springer Science+Business Media

springeronline.com

© Springer-Verlag Berlin Heidelberg 2005
Printed in The Netherlands

Cover design: KünkelLopka, Heidelberg
Typesetting: DA-TeX · Gerd Blumenstein · www.da-tex.de
Production: LE-TeX Jelonek, Schmidt & Vöckler GbR, Leipzig
Printed on acid-free paper 33/3142/YL - 5 4 3 2 1 0

Introduction

During the 1990's, the competitive pressures and short product lifecycles have caused many manufacturing and retail companies to focus on supply chain management practices and applications. Along with the Internet-driven e-Commerce, supply chain software companies became the darling of investors and supply chain practitioners. Indeed, more than any other three letter acronym initiatives such as MRP (Materials Requirements Planning), TQM (Total Quality Management), JIT (Just-In-Time), or CFM (Continuous Flow Manufacturing), Supply Chain Management (SCM) was a program which seemed to broaden the boundaries of business optimization beyond the four walls of the companies, as it addressed cross-organizational or even cross-company issues.

More demanding and sophisticated customers expected customized products (e.g. computers, cars) and short delivery times. Customers' buying decisions were often based on availability, not just on quality and price of the product alone. This created demand for more reliable, capacitated production planning and "available to promise" functionalities. Shorter product cycles not only changed the way products had to be manufactured, but also the way inventory (parts, sub-assemblies, and finished goods) had to be managed. Companies made big efforts to reduce their inventories, even with slogans such as "Zero Inventory", which of course was a vision that could not be achieved. However, various intelligent methods of inventory management were introduced to find optimal stock levels.

Continuing shifts in the geopolitical situation (e.g. NAFTA, Extension of European Union) and emerging markets (e.g. Eastern Europe and China) opened new business opportunities and at the same time kept companies busy revising their supply chain structures such as manufacturing locations, warehouse locations, inbound logistics, and distribution operations. This led to an increased demand in strategic supply chain planning tools such as supply chain simulators and location optimization tools.

Furthermore, companies have understood that in order to be more competitive, partners in a supply chain have to closely work together. However, mistrusts between partners have prevented them from adapting new available technologies to collaborate. Often the benefits of improved supply chain management practices have gone to the gorillas in the supply chain such as Wal-Mart and Dell, and much less to their smaller partners.

Indeed, competitive pressures of the 1900's have only gotten exacerbated, with global competition further squeezing profit margins, and the uncertain worldwide political and economic conditions have made the supply chain risks that much worse. Companies need to be able to react to changes more quickly and are also seeking new ways of avoiding risks or sharing risks with their supply chain partners.

As a result of the economic realities and understanding of supply chain management practices, many companies have introduced sound SCM practices and solutions (often referred to as APS or Advanced Planning and Scheduling). In addition most ERP (Enterprise Resource Planning) applications have now incorporated supply chain management functions and are becoming more mature, for example, SAP's APO modules. With the adoption of the Internet for businesses, some companies are successfully practicing collaboration over the Internet. With all these advances, yet, companies are looking for more differentiation to be competitive.

New techniques and practices for highly efficient supply chain management are being made possible by the rapid progress in information and communication technologies, laying the foundation for a new wave of applications. As we experience daily, performances of computer systems are still increasing exponentially. This includes processor speed as well as memory size and bandwidth. This enormous progress makes applications possible that were unthinkable a few years ago.

These advances are especially beneficial for quantitative models for decision making. Operations research methods such as mathematical programming, queuing and inventory theory, and stochastic optimization are receiving new attention in SCM applications, even though they had been in practice in other businesses for many years.

On the 'sell-side', the move from printed price lists and catalogs towards online price information communicated over the Internet has opened possibilities for more flexible pricing. With this, the manufacturing industry is adopting practices that e.g. the travel industry (e.g. airlines) has been using already for many years. On the 'buy-side', electronic connectivity between manufacturers and their suppliers opens new ways for negotiations and contract management. With new forms of more flexible contracts, both manufacturers and their suppliers can better cope with the uncertainties in demand, and its associated risks. Flexible contracts require new decision support systems that use stochastic versus purely deterministic techniques. These techniques have been practiced in the financial markets or the energy markets and now are being adopted for SCM practices. In manufacturing, production planning was traditionally done under deterministic assumptions. But the constraints in demand and supply, the manufacturing and transportation times, and the availability of resources are often stochastic in their nature. New approaches to cope with uncertainty in production planning have been demonstrated. An example is implosion technology that complements classical MRP bill-of-material explosion to take unforeseen material shortages into consideration.

The Internet has become a communication medium that is accessible from practically anywhere in the world. It exceeds the possibilities of electronic data interchange (EDI) by far. For a short period of time, between 1999 and 2001, this development was discussed with much hype under the notion of e-marketplaces. This idea which was driven by the dot.com hype was rather unproductive and therefore we avoid the term e-marketplaces. The fact is that the number of electronic transactions between companies is irresistibly increasing and the integration of processes and information are becoming more and more prevalent. During the past few years, companies have focused on integrating internal applications and business processes in order to reduce costs by automating many of the manual processes, including those associated with supply chain. The same companies have also been integrating with their external partners, suppliers and customers; however, much of such integration still has been in the form of EDI or through extranets which require manual entries. With the advent of new Internet-based standards such as XML and web-services, these companies should be able to integrate the supply chain processes with the external partners more flexibly and automatically. Business process standards such as RosettaNet for the electronics industry or CIDX for the chemical industry, in conjunction with the web-services technologies will also make partner integration much less expensive and time consuming. Even as we write this introduction, these industry organizations are actively working on the standards with the explicit goal of reducing the supply chain process integration costs.

With both increased computing power and connectivity, new applications come into reach, which extend the scope of decision making from single enterprise to multiple enterprises or even the entire supply chain network and which recognize the fact that decisions that have to be made within one enterprise cannot neglect the facts that are not controlled by themselves but are determined by the business environment such as changing demand, changing prices, or changing supply situations. Connectivity and the integration of business processes have laid the foundations for an increased visibility over the entire supply chain. Software technologies like portals, data warehouses, reporting systems, and on-line analytical processing (OLAP) are providing the necessary information and visibility for the decision makers. The next step is to assist the decision maker to quickly and optimally respond to unexpected situations. Intelligent analytics can automatically determine the best decision alternative and predict its consequence on the supply chain performance. We use the phrase "sense and respond" to characterize such a supply chain management system that is able to respond quickly and optimally to unexpected situations.

With the innovative practices and technologies described in this book, companies are able to reach a new level of excellence in managing their supply chains. We call this Supply Chain Management on Demand. According to IBM's definition, an On Demand Business is an enterprise whose business processes – integrated end-to-end across the company and with key partners, suppliers and customers – can respond with flexibility and speed to any customer demand, market opportunity or threat. An On Demand Supply Chain is a highly dynamic, adaptable business model that integrates information and decisions across all participants in an extended enterprise. Supply Chain Management on Demand is radically changing the way an company thinks about its organization and processes. It is transforming the supply chain from a competitive necessity to a competitive advantage.

In the remaining section of this introduction, we briefly review each of the chapters in this book. These chapters were written by supply chain researchers, consultants, and supply chain practitioners who have not only developed the practices but have deployed these practices in various supply chains at IBM and other companies. They address some of the advances in supply chain management practices we discussed above.

In Chapter 1, William Grey, Kaan Katircioglu, Dailun Shi, Sugato Bagchi, Guillermo Gallego, Mark Adelhelm, Dave Seybold and Stavros Stefanis present a new approach for rationalizing supply chain investments. Traditional ROI analysis has a number of shortcomings. Projected supply chain benefits, such as reductions in inventory carrying costs or logistics expenses, are notoriously difficult to quantify. Putting too much emphasis on cost savings and revenue improvements often means neglecting metrics that support long-term strategic objectives. And traditional approaches rarely consider risk. The analytic tools developed by Grey et al. overcome these shortcomings. These tools form the basis for new risk and opportunity assessments that consultants can use to help their clients make better and more intelligent decisions and extract greater value from their supply chain initiatives.

In Chapter 2, Steve Buckley and Chae An discuss the value of simulation in the context of analysis, planning and control of supply chains. Supply chain simulation complements other analytical techniques such as spreadsheets and mathematical optimization. A particular strength of simulation is the ability to consider uncertainty that is found everywhere in the supply chain, for example in customer demand, lead times and supply availability. Although optimization under uncertainty is an important research topic, few commercial supply chain optimization tools already support it. As Buckley and An point out, supply chain simulation has become an easily accessible, easily usable and flexible technology to address a wide range of supply chain problems under uncertainty.

Chapter 3, by Feng Cheng, Markus Ettl, Grace Lin and David Yao, focuses on inventory as one of the main cost drivers in supply chain management. Inventory costs, which include price protection, financing, inventory write-downs (price erosion), and inventory write-offs (obsolescence) can have a tremendous impact on business performance. The complexity of today's end-to-end supply chains makes it a serious challenge to determine where to hold safety stock in order to minimize inventory costs and to provide a committed level of service to the final customer. Cheng et al. describe the development of analytical models for the optimal placement of safety stocks in multi-echelon supply chains that are subject to forecast, lead time, and attach-rate uncertainty. They focus on applications in high technology supply chains.

In Chapter 4, Aliza Heching and Ying Tat Leung describe how traditional pricing practices are changing in the era of e-business. They provide an overview of common pricing practices and the strategic and tactical pricing-related decisions faced by a seller of products. They describe key features offered by commercial pricing systems. Finally, Heching and Leung review some case studies which demonstrate the level of financial benefits that have been derived from implementation of price optimization systems. The case studies also serve to illustrate the typical first steps taken by businesses that wish to experiment with price optimization.

In Chapter 5, Brenda Dietrich, Daniel Connors, Thomas Ervolina, J.P. Fasano, Robin Lougee-Heimer and Robert J. Wittrock give an example of how limitations of traditional manufacturing resource planning (MRP/MRP II) systems can be overcome by supplementary mathematical planning systems. MRP systems break down the finished goods demand into material requirements according to the BOM structure (material "explosion"). The assumption is that all materials – either produced in-house or ordered from suppliers – will be available when needed for production. This is rarely the case due to the uncertainties that are inherent in supply chains. Dietrich et al. have developed an "implosion technology" that takes into account materials shortages and solves the "resource allocation problem" to optimally determine which end products should be produced under the limited material availability. Successful deployments have shown that this technology can significantly reduce the cycle time of the planning process and increase manufacturing efficiency.

In Chapter 6, Robert Guttman, Jayant Kalagnanam, Rakesh Mohan and Moninder Singh provide an overview of the various functions in sourcing and procurement. They provide a brief description of IT technologies and the mathematics that underlie these technologies, discuss the functionality offered in current commercial platforms, and provide a roadmap of future useful features.

In Chapter 7, Colin Kessinger and Heiko Pieper present a solution that incorporates risk and uncertainty into sourcing and procurement decisions. The solution is based on mathematical models that adapt and extend financial engineering techniques. As Kessinger and Pieper point out, a number of companies have already adopted this technology to proactively manage risk and flexibility in their supply chains. Their Supply Risk and Flexibility Management (SRFM) framework focuses on risk-adjusted sourcing costs, quantifying the performance of supply agreements (contracts) against a range forecast. A set of industry examples spanning the Automotive, Consumer Packaged Goods and High-Tech sector demonstrate the use and benefits of this approach.

In Chapter 8, Moritz Fleischmann, Jo van Nunen, Ben Gräve and Rainer Gapp review the field of reverse logistics. They discuss its opportunities and its challenges and indicate potential ways for companies to master them. They highlight what makes reverse logistics different from conventional supply chain processes, but also point out many similarities. Fleischmann et al. review key results from academic literature and complement them with illustrations of reverse logistics practice at IBM.

In Chapter 9, Michel Draper and Alex Suanet give an overview of recent developments in service parts logistics management. They make a comparison with traditional (finished product) supply chain management and describe the specifics in the service parts supply chain. The concepts are illustrated by examples. Draper and Suanet see rapid changes in the service logistics environment and include their vision on further developments in the near future.

In Chapter 10, Santhosh Kumaran and Kumar Bhaskaran position business process integration as one of the major enabling technologies for supply chain management. In today's business environment, there is an increasing demand for flexibility of IT solutions. In order to stay competitive, enterprises must be able to quickly respond to changing business conditions. Business Process Integration and Management (BPIM) constitutes a set of technologies that serve as the foundation for creating flexible and agile supply chain solutions. The authors present a vision for future supply chain management systems, identify the technical challenges in realizing this vision, and outline a solution leveraging BPIM technologies.

In Chapter 11, Chris Nøkkentved gives an overview of collaboration in the supply chain. He explains the evolution from supply chains to e-supply networks, driven by the growing business usage of the Internet. He describes the new competitive landscape of e-supply networks and discusses new forms of inter-company relationships. Nøkkentved distinguishes collaborative processes between manufacturers and customers, between manufacturers and suppliers, and between manufacturers and 3rd party logistics providers. He concludes his chapter with implementation considerations.

In Chapter 12, Steve Buckley, Markus Ettl, Grace Lin and Ko-Yang Wang introduce the Sense-and-Respond paradigm for intelligent business performance management. Sense-and-Respond is a new customer-centered approach that provides real-time responsiveness necessary for organizations to proactively manage their supply chain. Buckley et al. describe a Sense and Respond Value Net Optimization framework that continuously recognizes and transforms events of business processes, generates and provides access to current business performance indicators, and immediately triggers appropriate actions across the entire enterprise and beyond. Two pilot applications of the Sense-and-Respond framework are presented.

Chae An and Hansjörg Fromm

Table of Contents

1 Beyond ROI[1]

William Grey, Kaan Katircioglu, Dailun Shi, Sugato Bagchi,
Guillermo Gallego, Mark Adelhelm, Dave Seybold, and Stavros Stefanis

Faced with heightened competition and a weak economy, companies are spending far more time developing business cases to justify their supply chain initiatives. Executives, consultants, software vendors, and project leaders alike have turned to return on investment (ROI) analysis as their tool of choice. But is this newfound interest in financial analysis paying off? Or is it just creating more confusion and sometimes driving poor investment choices? When it comes to analyzing supply chain initiatives, ROI analysis often falls short.

IBM Research, in collaboration with IBM Global Services, has developed a new approach for rationalizing supply chain investments. By taking ROI to the next level, it helps you make better decisions and extract greater value from your supply chain.

1.1 Where ROI Falls Short

When properly applied, ROI analysis is a powerful tool. And greater management attention to quantifying business impact certainly leads to more intelligent supply chain investments. However, ROI analysis is especially difficult to apply when analyzing supply chain improvements. Projected supply chain benefits, such as reductions in inventory carrying costs or logistics expenses, are notoriously difficult to quantify. Although the "hard" benefits reported in a typical ROI analysis may appear authoritative, they often rely heavily on assumptions about the impact of anticipated operational improvements. In many cases, these assumptions represent little

---

[1] Reprinted with permission of *Supply Chain Management Review*, Copyright 2003, Reed Business Information.

more than educated guesswork by individuals who won't actually be called on to deliver the improvements.

Another potential difficulty with traditional ROI analysis is that it doesn't consider interactions between supply chain initiatives. Firms typically have a portfolio of supply chain projects that they plan to deploy. Since these projects frequently have overlapping benefits, analyzing each project independently can lead to double counting. It may also ignore synergies between initiatives with interdependencies that make them *more* valuable when they are deployed together. Furthermore, unless a consistent framework is used to quantify the supply chain benefits of different projects, comparisons will not be valid.

The strength of ROI – its unerring focus on a narrow set of financial benefits – can also be a weakness. By placing too much emphasis on the cost savings or revenue improvements associated with a supply chain investment, companies may neglect other metrics that are also critical to business success. This encourages investments in initiatives that generate short-term gains, without helping the company achieve its long-term strategic objectives.

Another shortcoming of traditional ROI analysis is that it often doesn't adequately address risk. Initiatives may be delayed, or may not deliver their anticipated value. Projected benefits are sometimes sensitive to assumptions about the external business and economic environment. Initiatives that perform well in a strong market may deliver only limited value if economic conditions deteriorate. Other initiatives that position a company to take advantage of new market opportunities may show few tangible benefits when analyzed using static ROI analysis.

ROI analysis tends to be reactive, rather than proactive. Detailed business cases are usually created late in the game – *after* potential supply chain initiatives have received internal sponsorship and support. Instead of designing initiatives to improve their business, decision-makers find themselves designing business cases to justify their initiatives. By failing to act early, executives miss an opportunity to reshape or redirect their initiatives' scope or focus in ways that would increase business impact.

1.2   ROI is dead. Long live ROI ...

Despite its shortcomings, ROI analysis *can* and *does* provide a solid basis for analyzing business value. However, organizations often fail to exploit its full potential. By carefully rethinking your approach to the ROI process, you can transform ROI into a far more effective decision-support tool. To help you do this, we recommend six concrete steps you can take to extract additional value from your supply chain investments:

- *Go deeper*. Analyze the causes of supply chain value, not just the effects.
- *Quantify the impact*. Build a richer model to evaluate the link between supply chain performance and business value.
- *Be consistent*. Develop a common, consistent framework for comparing and evaluating initiatives.
- *Don't just follow the money*. Consider management objectives that go beyond immediate financial benefits, and focus on strategic intent.
- *Consider risk*. Understand each initiative's likelihood of success, and how it helps your supply chain adapt and respond to changes in your business environment.
- *Put ROI to work*. Don't just use financial analysis to defend your initiatives. Instead, use it to define your supply chain strategy, and to focus and manage your supply chain efforts.

Properly applied, these steps can help you make better investment decisions, thus improving the performance of your supply chain.

1.2.1 Action 1: Analyze Supply Chain Value Drivers

Traditional ROI analysis works well when quantifying the benefits of a supply chain initiative that has a direct impact on *financial* performance. This would be the case, for example, for an initiative focused exclusively on cutting costs through headcount reductions.

In practice, though, supply chain initiatives usually deliver much of their business value by improving *operational* performance. Such improvements ultimately translate into better financial performance. However, estimating *how much* they improve financial performance requires going beyond traditional ROI.

Consider, for example, the case of a hypothetical personal computer manufacturer considering a major redesign of its supply chain. The company's management was evaluating a number of initiatives, including demand planning, supply network planning, and order fulfillment management solutions.

The first initiative being evaluated was the demand planning solution. By automating the planning process, it would lead to a small cut in headcount. But the real payoff was expected to come from significant reductions in Finished Goods Inventory. The company's management team was comfortable estimating savings from headcount reductions. But how much would inventory *really* come down? Without a credible mechanism for evaluating the financial impact of operational improvements, management would be forced to resort to educated guesswork.

The first step in quantifying the impact of the demand planning solution is to understand how it affects the company's *supply chain value drivers*. By our definition, a supply chain value driver is an operational metric that passes two important tests. First, it must be directly affected by a supply chain solution or initiative. Second, the metric must have an impact – albeit an indirect one – on at least one of the firm's Key Performance Indicators (KPIs). Think of a supply chain value driver as a "lever" that an initiative can turn to impact business performance.

The proposed demand planning solution affects a number of supply chain value drivers, including demand planning cycle time and forecast accuracy (see Fig. 1.1) These supply chain value drivers (and the others shown in the figure) ultimately impact one of the firm's KPIs – in this case Finished Goods Inventory. For example, shorter planning cycle times make a business more responsive to variability in customer demand, reducing the need for inventory. And more accurate demand forecasts make it easier to effectively match production to customer demand, also resulting in less inventory.

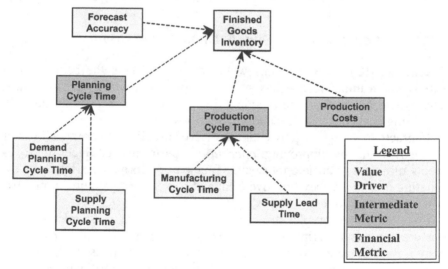

Fig. 1.1. Inventory Value Drives

Since the impact of an initiative on supply chain value drivers is relatively easy to estimate, value drivers provide a solid basis for assessing business impact. The management team evaluating the demand planning solution was having a difficult time determining how much the initiative would affect inventory levels. However, they were reasonably certain of its impact on key value drivers: The initiative would cut weekly demand planning cycle times in half, and increase forecast accuracy by about 10 percent.

1.2.2   Action 2: Quantify Value Driver Impact

Of course, understanding the relationships between supply chain value drivers and financial performance is only half the story. The next step is *quantifying* these relationships. There are a number of ways to accomplish this, including interviews with subject matter experts, analysis of historical data, pilot projects, and mathematical modeling.

The simplest approach is to interview subject matter experts and solicit their estimates of how much a change in a value driver, such as forecast accuracy, would affect inventory levels. Although this approach involves more art than science, it still increases the accuracy of the ROI analysis.

An even more effective approach is to analyze historical data. By comparing historical changes in value drivers such as forecast accuracy with changes in financial metrics like inventory levels, a clearer picture of the quantitative relationship between the two begins to emerge.

One frequent problem with this approach is the difficulty of finding accurate data. Even if data is available, analyzing it can be tricky. Many factors besides forecast accuracy affect inventory levels, and it may be difficult to determine their relative contributions. If a firm's operating environment changes, old relationships may no longer be valid.

Techniques such as multiple regression analysis can be used to assess the impact of supply chain value drivers, such as forecast accuracy, on inventory performance while controlling for other variables. If benchmarking data is available, it can be analyzed in a similar fashion. However, many companies are skeptical of using benchmarking data, because they believe that it may not reflect the unique characteristics of their business.

An excellent way to quantify the relationship between value drivers and financial metrics is to use data collected during a pilot project. A pilot is a project of limited scope, designed to probe the potential of a major initiative before committing to its deployment. Pilots can be carefully controlled and monitored, so data availability is usually not a problem. Because a typical pilot impacts only a few value drivers and financial metrics, analysis of the pilot results is comparatively straightforward. Results from the pilot also reflect current operating conditions. Despite the benefits, in many cases it is not feasible to perform a pilot, because it may introduce additional costs or unacceptable delays.

Finally, the impact of supply chain drivers can be quantified using mathematical models. For example, IBM has created its own proprietary model, which it has used successfully for evaluating supply chain initiatives. This approach has a number of distinct advantages. Once a model has been built and validated, it can be broadly applied to different situations, even if they have not been encountered in the past. Complex interactions between supply chain value drivers and financial metrics are often difficult to estimate empirically; without using a mathematical model it may not be possible to effectively capture such interactions. Over the last several decades, a broad range of supply chain issues have been addressed in management science and operations research. Practitioners have developed a number of stochastic quantitative models that can be applied to estimate the impact of many supply chain value drivers on certain key financial metrics. In addition, techniques such as Monte Carlo simulation and discrete event simulation can provide further insights into the link between operational changes and financial performance (see section 1.4: Supply Chain Value Modeling at IBM).

To illustrate this approach, we used a mathematical model to analyze the impact of several key supply chain value drivers on a number of financial and operational metrics (see Figure 1.2). The figure shows the impact of each value driver on one of these metrics – finished goods inventory. For the drivers analyzed here, inventory was most sensitive to changes in forecast error, supply lead time, production planning cycle time, and master planning cycle time. Sensitivity was measured as the inventory improvement caused by a unit change in the value driver. For example, the figure shows that a one-day reduction in supply lead time would result in about a three million dollar inventory improvement.

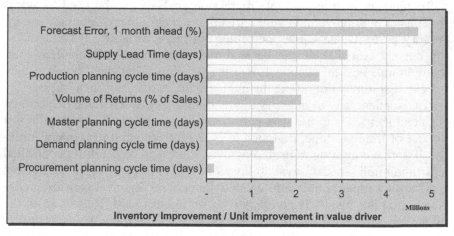

Fig. 1.2. Financial and Operational Impact of Key Value Drives

1.2.3   Action 3: Use a Common Framework

Typically when organizations evaluate a portfolio of supply chain initiatives, they create a separate business case for each initiative. Each business case has a unique set of assumptions about how its initiative will affect financial performance. Initiatives are analyzed independently, with no consideration of how interactions between initiatives affect performance.

One of the biggest advantages of quantifying impact at the value driver level is that it provides a common basis for evaluating multiple supply chain initiatives. Instead of being viewed in isolation, these projects can be analyzed as a portfolio. A single model is used to evaluate each initiative. The model uses a common set of assumptions about the impact of supply chain value drivers on financial performance. This helps create a level playing field for comparing initiatives. It also provides a framework for using more sophisticated techniques to analyze initiative interactions, such as synergies and diminishing returns. (See section 1.5: Understanding Value Driver Interactions).

To illustrate this approach, we present a simplified example of how it could be applied to estimate the business impact of a set of initiatives. Although our example focuses on calculating inventory impact, in practice other measures of financial performance would be considered as well.

The process begins by estimating the impact of each initiative on key supply chain value drivers (see Fig. 1.3, which shows a table with the output of this step for the hypothetical personal computer manufacturer). The first column in the table has a list of supply chain value drivers. The next three columns show how each initiative is expected to affect each value driver.

Value Drivers	Impact of Solution on Value Drivers			Impact of Value Driver on Inventory
	Demand Planning Solution	Supply Network Planning Solution	Order fullfillment management Solution	
Procurement planning cycle time (days)	-	23	-	200,000
Master planning cycle time (days)	3.5	4	-	1,900,000
Supply Lead Time (days)	-	6	-	3,100,000
Production planning cycle time (days)	3.5	6	-	2,500,000
Premium Freight Costs($)	-	$ 420	-	-
Demand planning cycle time (days)	3.5	-	-	1,500,000
Forecast Error (%)	10%	-	-	4,700,000
Customer Service Management Costs ($)	$ -	$ -	$ 721	-
Defective Orders (% of total orders)	-	-	0.5%	-
Invoicing Errors (% of total invoices)	-	-	2.3%	-
Total order processing time (days)	-	-	4.3	-
Volume of Returns (% of Sales)	-	-	0.5%	2,100,000
Order ship to Customer Invoice (days)	-	-	11.0	

Fig. 1.3. Impact of Solutions on Key Value Drivers

The proposed demand planning solution supports collaborative forecasting and planning with customers. It is thus expected to improve forecast accuracy and enable the company to move from weekly to bi-weekly planning for a number of key processes. These expected improvements are entered in the column labeled "demand planning solution." As shown in the table, the solution is expected to cut each of these planning cycle times by three and half days, and reduce forecast error by 10%.

Similar entries are made for the other two solutions. Supply network planning is expected to reduce procurement planning cycle times from a month to a week and enable the company to perform daily production scheduling. It would also dramatically shrink supply lead times and cut costs for expediting component shipments. Order fulfillment management would help the company increase the speed and accuracy of its order processing, shipping, and invoicing processes.

The inventory impact of each solution can now be calculated. To make this easier, we included a column in the table showing the impact of each value driver on finished goods inventory. (Note that this is the same information that was shown graphically in Fig. 1.2.)

According to the table in Fig. 1.3, the demand network planning solution is expected to reduce master planning cycle time by three and a half days. For each day the company reduces master planning cycle time, it is projected to cut inventory by about $1.9 million. By taking the product of these numbers, we estimate that master planning cycle time reductions would reduce inventory by $6.65 million. ($1.9 million times 3.5.)

Of course other value drivers, such as procurement planning cycle times and production planning cycle times would be expected to provide additional inventory benefits. And this simple calculation ignores issues like diminishing returns and synergies. More sophisticated modeling techniques, however, can be used to analyze these complex value driver interactions.

Fig. 1.4 shows model outputs reporting the annual financial benefit of each solution, after accounting for the impact of multiple drivers and value driver interactions. Supply network planning has by far the greatest impact, followed by demand planning. Order fulfillment management is a distant third. For all solutions, the most significant savings were associated with Inventory reductions. These included savings in inventory carrying costs and a reduction in write-downs for inventory obsolescence and price declines.

1.2.4   Action 4: Link ROI to Business Strategy

One shortcoming of traditional ROI is that it places too much emphasis on a narrow set of financial benefits. This encourages investments in supply chain initiatives that generate short-term gains but fail to cultivate the key resources and capabilities needed to achieve long-term strategic objectives.

To overcome ROI's myopic focus, extend your analysis to consider a broader set of metrics. This can be done formally, by linking the analysis directly to a balanced scorecard that includes both financial and non-financial metrics. Or it can be done informally, by identifying a set of KPIs considered critical to business success. The financial benefits reported in a traditional ROI analysis can then be extended to include these additional KPIs. Not all balanced scorecard metrics can be readily quantified, of course. However, including those that *can* improves decision making by providing a clearer view of business impact.

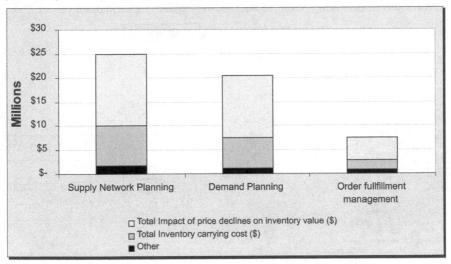

Fig. 1.4. Financial Impact

The hypothetical personal computer manufacturer tracked a number of KPIs, including net income, inventory turnover, and on-time delivery. All three measures were an integral part of the firm's balanced scorecard: net income as a financial metric, inventory turnover as a measure of the effectiveness of internal business processes, and on-time delivery to assess performance from a customer perspective.

Net income was already reported as part of the firm's ROI process. However, inventory turnover and on-time delivery were not. To calculate these additional balanced scorecard measures, follow an approach similar to value-driver analysis.

The first step is to analyze the relationships between the additional KPIs and the financial and operational metrics quantified earlier (see Fig. 1.5). In some cases, these relationships are simply accounting identities.[2] For example, inventory turnover is the ratio of cost of goods sold (COGS) to inventory.[3] In other cases, supply chain value drivers need to be analyzed

---

[2]   For a good treatment of accounting ratios, see: Stickney (1998).

[3]   Strictly speaking, inventory turnover is usually defined as cost of goods sold divided by *average* inventories.

as well. For example, on-time delivery performance depends on the balance between finished goods inventory levels and end-to-end supply chain cycle time. Companies with long end-to-end cycle times are less responsive to shifts in customer demand and can only achieve high on-time delivery performance by holding additional inventory. More responsive firms can hold less inventory without sacrificing delivery performance.

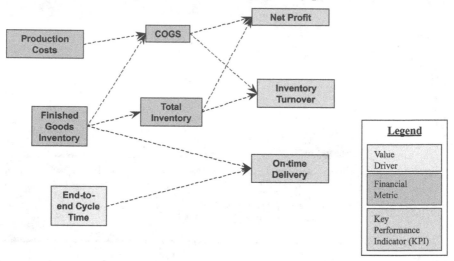

Fig. 1.5. Linking ROI to Strategic Metrics

The next step is to quantify the relationships. For accounting ratios such as inventory turnover, this just requires a simple calculation. More sophisticated analysis is needed to quantify complex interactions – such as the relationship between inventory, supply chain responsiveness, and on-time delivery. Again this can be addressed in a number of ways, including interviews with subject matter experts, historical analysis, and mathematical modeling.

Analyzing a broad set of strategic metrics can provide additional insights that improve decision making. To illustrate this, we calculated the impact of each solution on a number of the personal computer manufacturer's balanced scorecard metrics (see Fig. 1.6).

When viewed through a purely financial lens, supply network planning is by far the most attractive solution (see Fig. 1.4). Its impact – in terms of the narrow measure of financial return reported in a typical ROI analysis – was over twice that of the order fulfillment management solution.

However, when viewed from a strategic perspective, a different story emerges. Order fulfillment management looks much stronger. Because of its impact on accounts receivables, the solution significantly improves on-time delivery, an important contributor to customer satisfaction. It also affects multiple financial metrics, including shareholder value added, cash-to-cash cycle time, and days receivables outstanding. Of the three solutions, it has the broadest and most significant impact on the firm's balanced scorecard. Strategically, order fulfillment management appears to be a more compelling choice.

Key Performance Indicators	Base Case	Supply Network Planning	Demand Planning	Order fulfillment management
Shareholder Value Added (EVA)	O	O	O	▶
Cash to Cash Cycle Time	O	◀	◀	●
Inventory Turns	O	●	●	O
On Time Delivery	O	O	O	●
Net Income	O	O	O	O
Receivables Outstanding (days)	O	O	O	●
End-to-end cycle time (days)	O	◀	●	O
Erroneous Invoice (%)	O	O	O	●

From	To	Key
	-20%	●
-20%	-10%	◀
-10%	10%	O
10%	20%	▶
20%		●

Fig. 1.6. Impact of Solutions on Strategic Metrics

1.2.5   Action 5: Consider the Risks

Traditional ROI analysis often fails to adequately address risk. For supply chain initiatives, risks can be grouped into two broad categories: implementation risks and business risks. Implementation risks include project delays, cost overruns, and outright project cancellations. Business risks are changes in the business or operating environment that either render an initiative obsolete or impair its ability to deliver business value.

The first step in analyzing implementation risk is to develop a qualitative understanding of the levels of risk associated with different initiatives. In general, initiatives with broader scope and greater complexity have higher risk. For IT projects, the maturity of the technology also affects risk. For example, it is generally less risky to implement packaged software than to develop a custom application. Projects can be ranked according to their perceived level of risk. Decision makers can assign a higher hurdle rate to projects with greater risk or insist on a more rapid payback (Hubbard 1998).

To gain a better understanding of business risks, begin by asking yourself what can go wrong. Then consider how changes in the business environment will affect results. Will the solution still deliver its anticipated benefits even if business conditions deteriorate? What potential problems will emerge if conditions suddenly improve?

When assessing risk, don't forget to consider the upside. With supply chain investments, you are often paying for increased efficiency, flexibility, and responsiveness. Increased speed and flexibility creates opportunities to gain market share if industry conditions unexpectedly improve. And initiatives intended to improve supply chain efficiency may provide a valuable safety net during an industry downturn.

Scenario analysis works well for quantifying how different initiatives respond to risk – both on the upside and the downside. Scenario analysis begins by defining a set of "what-if" scenarios that correspond to risky future states of the world. Each scenario modifies key assumptions of the original business case to reflect changing operational, business or economic conditions. A new business case is generated for each scenario, which is then compared with the original. Companies can develop a common set of risk scenarios and then use them to test each initiative being evaluated.

To show the insights that can be gained by assessing risk, we used scenario analysis to test how two solutions being considered by the personal computer manufacturer – demand planning and order fulfillment management -- would perform under different business conditions. We considered two scenarios. In the soft demand case, we assumed that a tough business climate caused a sudden decline in unit sales, accompanied by pricing pressure. In the strong demand case, we assumed that improving economic conditions increased unit sales and firmed up pricing.

Notice how differently business risk affects the two solutions (see Fig. 1.7). When demand is soft, the demand planning solution actually performs better. By improving forecast accuracy and responsiveness, the solution enables the company to more rapidly adjust its inventories to match lower levels of customer demand. When compared to its performance in the base case, it delivers an additional two million dollars in annual benefits. Order fulfillment management, on the other hand, performs worse in a down market. Although the solution improves order management execution by increasing the speed and accuracy of order management processes, it does nothing to improve supply chain responsiveness. When demand drops, it provides no protection against a build-up of excess inventory.

The results are similar when demand rises unexpectedly. The demand planning solution enables the personal computer manufacturer to perform more effective inventory planning, thus reducing lost sales due to inventory stock-outs. Although order fulfillment management improves execution, it can't prevent an increase in lost sales.

In the previous section, we found that the Order fulfillment management solution had a significant impact on a broad set of KPIs. However, in the highly volatile personal computer industry, the Demand Planning Solution does better at reducing risk.

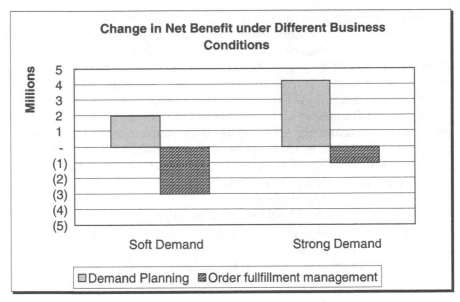

Fig. 1.7. Risk Analysis: Change in Net Benefit under Different Business Conditions

1.2.6 Action 6: Put ROI to Work

Most supply chain organizations take a just-in-time approach to ROI. They dust off the spreadsheets and crank the numbers when the time comes to develop a financial justification. Once a project has been approved, financial analysis becomes a thing of the past. However, the focus on business value should not begin and end with the business case. Once you have developed an effective and consistent framework for analyzing business impact, there are a number of ways you can use it to improve decision making.

First, begin to apply financial analysis earlier, *before* you have developed your supply chain strategy. You can eliminate much of the guesswork from the strategy process by identifying the supply chain value drivers with the greatest impact on your company's business objectives and using this knowledge to specifically target initiatives that impact these drivers. Mathematical models are particularly effective at this stage because they can be used as a diagnostic tool to analyze the unique characteristics of your business and then to pinpoint opportunities for improvement. Mathematical models also make it easy to rapidly analyze and test multiple options using what-if analysis.

Financial analysis can also play an important role *during* project development and rollout. An analysis of the project's potential business impact can help you make more intelligent choices when defining project scope and functionality. By understanding how interactions between initiatives affect business performance, you can do a better job at sequencing and prioritizing the rollout of your initiatives. Additional analysis can also help identify opportunities to redefine and redirect your supply chain efforts when business conditions change or project schedules start to shift.

Finally, our framework can help monitor the performance of initiatives *after* they have been deployed. Assessing the performance of an initiative by monitoring its performance against financial objectives is often a challenge. Because so many factors can affect financial measurements, they often behave like a moving target. This can make it difficult to judge how well a project is meeting its objectives.

This problem can be addressed by using supply chain value drivers – rather than financial metrics – as performance benchmarks. Because initiatives act directly on value drivers, their performance can be more easily measured. It thus becomes easier to tell whether a project is actually delivering value. The result: less finger pointing, and more accountability.

1.3 Making it Happen

The transition to an integrated process for managing your portfolio of supply chain initiatives will not take place overnight. To ease the transition, we suggest you begin with an incremental approach. Start by analyzing a single project. This gives you an opportunity to learn more about your supply chain value drivers and to quantify their impact. It also enables you to begin developing an integrated framework for analyzing performance and risk.

To ensure success, choose your first project carefully. This initial project should be highly visible and ideally have CEO sponsorship. As part of the project's budget, funds should be specifically allocated to develop the framework. Make sure that the CFO supports the effort and gets his staff involved.

Once you have begun to develop the framework, build on it. As new initiatives are conceived, evaluate them using the framework. Begin to consider interactions between new initiatives and projects that are already underway. Eventually, you can use the framework to guide the design, timing, and scale of all new initiatives.

Most companies base their supply chain decisions on past trends and experience. As a result, they often find themselves fighting last year's battle, only to fall further behind their best-in-class competitors. Making the right decisions requires a deeper understanding of how changes in your supply chain impact business performance. By using a consistent technique for selecting and managing your supply chain initiatives, you can make the hard investment choices while still investing to stay ahead of the competition.

1.4 Supply Chain Value Modeling at IBM

Over the last year, IBM Research has been working closely with IBM Business Consulting Services to develop analytic tools to quantify the business value associated with Supply Chain initiatives. These tools form the basis for a new risk and opportunity assessment that IBM is using to help its clients evaluate supply chain initiatives. This work has been applied in a number of industries, including Automotive and Electronics, and for a number of applications, including complex manufacturing and distribution.

The model's analytic approach is based on two key streams of applied research. The first is an IBM Research effort that utilizes Management Science and Operations Research modeling techniques to develop a deeper understanding of the impact of operational drivers on an enterprise's operating performance. The second is a cross-disciplinary IBM Research effort that integrates tools and techniques from the domains of Finance and Supply Chain Management to improve overall business performance. These two streams of research have been applied to improve financial and operating performance at a number of IBM divisions, including IBM's Server, Personal Computer, Microelectronics, and Storage business units (Lin et a. 2000).

The model estimates the impact of factors such as demand variability, supply delays and production cycle times on metrics such as inventory and customer service. This is accomplished using stochastic quantitative models to quantify the link between changes in physical, temporal and informational flows and financial and operational performance.

Supply chain processes modeled in detail include planning, procurement, production, and order management. These processes affect value chain performance by introducing delays, variability, and constraints into the system. Delays make the enterprise less responsive to changes in the external business environment. Variability influences the consistency of value chain performance, making it harder to meet customer service targets. Constraints, such as limited production capacity, affect the magnitude of the enterprise's response to change.

1.5   Understanding Value Driver Interactions

When evaluating initiatives that touch on related supply chain value drivers, it is important to consider diminishing returns and synergies. Theoretical models can be especially useful for untangling these and other complex value driver interactions.

Diminishing returns occur when benefits taper off with increasing improvements to a supply chain value driver. For example, when forecast accuracy is extremely poor, a small forecasting improvement has a big inventory impact. However, when forecast accuracy is high to begin with, a similar change results in a much smaller improvement.

For value chain drivers with diminishing returns, organizations realize the greatest impact by attacking metrics with the poorest performance. It is important to consider diminishing returns when an initiative has a major impact on a value driver, or when multiple initiatives affect the same driver.

Synergies are another important interaction effect. Synergies occur when the combined impact of a group of value drivers working together is greater than the total impact of each working separately. Understanding value driver synergies helps management deliver additional value by targeting initiatives that work especially well together.

References

Stickney, C.P. (1998) *Financial Reporting and Statement Analysis*, International Thomson Publishing
Hubbard, D. (1998) Hurdling Risk, *CIO Magazine*, June 15
Lin, G., Ettl, M., Buckley, S., Bagchi, S., Yao, D.D., Naccarato, B.L., Allan, R., Kim, K., Koenig, L. (2000) Extended-Enterprise Supply-Chain Management at IBM Personal Systems Group and Other Divisions. *Interfaces* 30 (1): 7–21

2 Supply Chain Simulation

Steve Buckley and Chae An

2.1 Introduction

Analysis, planning and control of a supply chain calls for a combination of spreadsheet, optimization and simulation models. Spreadsheet analysis is by far the most popular form of supply chain modeling due to its accessibility, ease of use and flexibility. However, spreadsheets are fairly limited in modeling power, with a few notable exceptions (Katircioglu et al. 2002). Optimization technology such as linear or mixed integer programming is a great way to solve well-defined mathematical problems such as supply network planning and inventory optimization (Ettl et al. 2000). But optimization models are rigidly structured and often based on simplifying assumptions to make the problem fit the mathematical format required by the underlying solver. Another issue that often limits the utility of optimization is uncertainty. Uncertainty abounds in supply chains – for example in customer demand, lead times and supply availability. Although optimization under uncertainty is a popular research topic, few commercial supply chain optimization tools support uncertainty models.

Simulation is a popular alternative to optimization for supply chain analysis. Simulation models are not restricted by rigid mathematical structures. Almost any supply chain issue can be coded as a simulation object with a set of parameterized behaviors. Individual components of a supply chain can be modeled separately and then combined into one large simulation model to represent the overall system. With simulation it is relatively easy to incorporate uncertainty, by generating random numbers for uncertain parameters during simulation runs. Multiple iterations are required to understand the results of uncertainty.

To be fair, it should be noted that simulation is an evaluative technique and does not automatically produce an optimal solution, unless simulation runs are controlled by an external search loop using an approach like Design of Experiments (Ermakov and Melas 1995). Moreover, supply chain simulations tend to be numerically intensive and sometimes take a long time to execute. Multiple iterations significantly increase the simulation runtime, as do external search loops such as Design of Experiments.

2.1.1 Comparison to Business Process Simulation

Off-the-shelf business process simulation tools have been readily available for about ten years. The popular commercial tools include ARIS (www.ids-scheer.com), Extend (www.imaginethatinc.com) and WBI Modeler (www.ibm.com/software/integration/wbimodeler). These simulators support general-purpose modeling of business activities but typically do not support detailed supply chain data structures such as demand forecasts and bills of material; supply chain algorithms such as production planning, forecasting and replenishment; or supply chain policies like build-to-plan and build-to-order. Cycle times and resource usage are the primary outputs of business process simulations. While supply chain simulations are also concerned with these generic outputs, they are also focused on more specific metrics like inventory and customer service. There is a saying in supply chain circles that any mistake will lead to excess inventory and lower customer service. Calculating inventory and customer service requires detailed numerical data structures, algorithms and policies that are not found in general-purpose business process models.

If business process simulation tools were truly extensible, they would make it possible for supply chain developers to create and share libraries of supply chain data structures, algorithms and policies. In some cases, modelers have built upon business process simulation tools to analyze detailed supply chain issues, albeit for onetime analyses (Feigin et al. 1996).

2.2 Simulation Modeling Requirements

In this section, we discuss the modeling requirements necessary for accurately analyzing supply chain issues through simulation technology. These requirements are derived from a large number of supply chain modeling studies that researchers at IBM have performed during the past ten years.

Plan
P1 Plan Supply Chain
P2 Plan Source
P3 Plan Make
P4 Plan Return

Source
S1 Source Stocked Product
S2 Source Make-to-Order Product
S3 Source Engineer-to-Order Product

Make
M1 Make-to-Stock
M2 Make-to-Order
M3 Engineer-to-Order

Deliver
D1 Deliver Stocked Product
D2 Deliver Make-to-Order Product
D3 Deliver Engineer-to-Order Product

Source Return
R1 Return Defective Product
R2 Return MRO Product
R3 Return Excess Product

Deliver Return
R1 Return Defective Product
R2 Return MRO Product
R3 Return Excess Product

Fig. 2.1. Fundamental management processes in the Supply-Chain Council's SCOR-model

2.2.1 Data

As mentioned in the previous section, general-purpose business process simulation tools typically do not support detailed supply chain data structures. Here are some examples of data structures that are often critical to supply chain modeling:

- **Product** definitions
- **Bills of Material** (BOMs) for product assembly
- **Customer Demand** for products

- **Customer Classes** describing customer service requirements for different types of users
- **Demand Forecasts** predicting customer demand over future time periods
- **Initial Inventory** levels for products at a location
- **Storage Space** definitions including storage size and associated costs
- **Reorder Points** for maintaining stock levels
- **Lot Sizes** for inventory replenishment
- **Supply Constraints** limiting the number of products available from an external supplier over a period of time
- **Locations** of customers, distribution centers and manufacturing sites
- **Routes** between locations and the transport time between them

2.2.2 Processes

The Supply-Chain Operations Reference (SCOR) model (Supply-Chain Council 2001) provides a starting point for building a simulation model of a supply chain (see Figure 2.1). The SCOR-model identifies five fundamental supply chain management processes: Plan, Source, Make, Deliver and Return. We have found it extremely useful to model these fundamental processes within the context of well-known supply chain business functions. Based on our experience, the following business functions are sufficient to model a variety of supply chain issues across many industries: Customer, Manufacturing, Distribution, Retail, Transportation, Inventory Planning, Forecasting and Supply Planning.

It is important to understand the scope of each fundamental process with respect to the business functions. The Plan process can apply to a single business function or to a set of business functions. For example, a Manufacturing function may plan only its own activities based on inputs it receives from other business functions in its supply chain. In other cases planning may be performed across business functions in an attempt to maximize overall supply chain value. For this reason three pure planning functions have been included in our list of business functions. The other fundamental processes – Source, Make, Deliver and Return – normally apply to only a single business function.

For modeling purposes one can parameterize each business function in terms of the fundamental processes it executes. The following descriptions provide a high-level overview of this parameterization:

- **Customer**. This business function represents end customers that issue orders to other business functions. Customer functions execute the fundamental processes Plan, Source and Return. Orders are generated on the basis of customer demand, which may be modeled as a sequence of specific customer orders (possibly obtained from historical records) or as aggregated demand over a period of time (that must be randomly disaggregated during a simulation run). The Customer function may also specify the desired due date, service level and priority for orders. Customer functions may send forecasts of future demand to other business functions.

- **Manufacturing**. This business function models assembly and maintains raw material and finished goods inventory. Manufacturing executes the fundamental processes Plan, Source, Make, Deliver and Return. Note that one Manufacturing function can supply another Manufacturing function, so there is no need to have a distinct function to model suppliers. A Manufacturing function makes use of modeled information such as the types of manufactured products, their manufacturing cycle time, bills of material, manufacturing and replenishment policies for components and finished goods, reorder points, storage capacity, manufacturing resources, material handling resources and order queuing policies.

- **Distribution**. This business function models distribution centers and warehouses, including finished goods inventory and material handling. Distribution functions execute the fundamental processes Plan, Source, Deliver and Return. A Distribution model typically includes inventory replenishment policies, reorder points, storage capacity, material handling resources and order queuing policies.

- **Retail**. This business function models retail stores, including finished goods inventory and material handling. Retail stores execute the fundamental processes Plan, Source, Deliver and Return. A Retail model typically includes inventory replenishment policies, safety stock policies, reorder points, material handling resources, backroom storage capacity and shelf space.

- **Transportation**. This business function models transportation types (e.g. trucks, planes, trains, boats), cycle time between shipping locations, vehicle loading and transportation costs. Transportation executes the fundamental processes Plan, Deliver and Return. A Transportation model typically includes order batching policies (by weight or volume), material handling resources and transportation resources.

- **Inventory Planning**. This business function models periodic setting of inventory target levels. Inventory Planning executes the fundamental process Plan. This business function may link to an optimization program that computes recommended inventory levels based on desired customer serviceability, product lead times and other considerations.

- **Forecasting**. This business function models product sales forecasts for future periods. Forecasting executes the fundamental process Plan. This business function may link to an optimization program.
- **Supply Planning**. This business function models bill-of-material explosion and allocation of production and distribution resources to forecasted demand under capacity and supply constraints. Supply Planning executes the fundamental process Plan. This business function may link to an optimization program.

In a supply chain it is important to distinguish between execution and planning processes. Execution processes are driven by plans and policies generated by planning processes. Both information and physical goods enter and leave execution processes. Planning processes deal only with information, not physical goods. Three of the business functions listed above are pure planning functions: Forecasting, Inventory Planning and Supply Planning. The other business functions can have a mixture of execution and planning processes.

2.2.3 Entities

In a simulation model, the items that enter and leave business processes are often referred to as *entities* or *artifacts*. Here is a list of entities that are specific to supply chain processes:

- **Request Orders** represent customer or replenishment orders for physical goods. These entities carry order information from Customers to Manufacturing and Distribution functions and from Manufacturing and Distribution functions to other Manufacturing and Distribution functions.
- **Filled Orders** represent customer or replenishment orders for which physical goods have been provided. These entities carry order physical goods from Manufacturing and Distribution functions to Customer, Manufacturing and Distribution functions. Filled orders may pass through Transportation functions where aggregation and transport occurs.
- **Shipments** represent a group of Filled Orders in transport. These entities carry Filled Orders from Transportation functions to Customer, Manufacturing and Distribution functions.
- **Forecasts** represent demand forecasts for customer and replenishment orders. These entities often carry demand forecast information from Forecasting functions to Supply Planning, Manufacturing and Distri-

bution functions. It is also possible for a Customer, Manufacturing, or Distribution function to have its own local forecasting process. If such a function shares its forecasts with other supply chain functions, it would do so by sending Forecast entities.

- **Supply Plans** represent production and procurement plans generated by a Supply Planning function, often based on forecast information. These entities usually carry information from Supply Planning functions to Distribution and Manufacturing functions.

2.2.4 Resources

The resource models provided by general-purpose business process simulators are often useful for supply chain simulation. Since cycle time and resource cost are key metrics in both business process and supply chain simulations, business process resource definitions can sometimes be reused for supply chain simulation. However, additional parameters and constructs are often needed to model the following supply chain resources:

- **Storage Resources** model cost and capacity of space where Manufacturing, Distribution and Transportation functions store physical goods.
- **Material Handling Resources** model cost and capacity of personnel and equipment used to move physical goods within Manufacturing, Distribution and Transportation functions.
- **Manufacturing Resources** model cost and capacity of personnel and equipment used to manufacture physical goods in Manufacturing functions.
- **Transportation Resources** model cost and capacity of vehicles such as trucks, trains and ships in Transportation functions.

2.2.5 Supply Chain Process Example

To illustrate a supply chain process, let us examine the Manufacturing function shown in Figure 2.2. A Manufacturing function assembles, stores and sells physical goods (products). A Manufacturing function can supply a Customer function, a Distribution function, or another Manufacturing function with products. It may require raw materials which can be ordered from a Distribution function or another Manufacturing function. Production schedules can be driven by Supply Plans received from Supply Planning functions or by Request Orders.

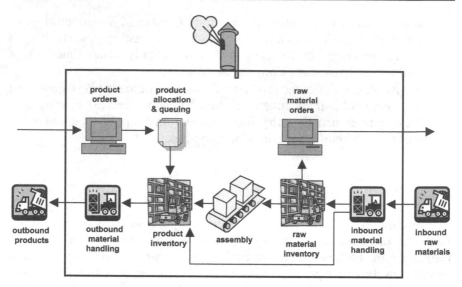

Fig. 2.2. Overview of a Manufacturing function

The main activities that take place inside a Manufacturing function are:

- Assembly
- Order queuing
- Production scheduling
- Product allocation
- Inventory replenishment
- Material handling

The raw materials needed for an assembly operation are based on the Bill of Material (BOM) definition for the product being assembled. A BOM lists the components and the quantities needed to assemble a unit of product. The length of time that elapses during an assembly depends on a specified manufacturing delay parameter. It also depends on whether the assembly is performed in parallel or sequentially. Under *parallel assembly*, all units of an order can be assembled at the same time. Under *sequential assembly*, only one unit can be assembled at a time. To illustrate, parallel assembly of an order for 5 computers with a 24 hour delay will be completed in 24 hours – as long as raw materials and Manufacturing resources for 5 computers are available. For sequential assembly, an order for 5 computers will take 120 hours to complete (5 times 24) – requiring raw materials for 5 computers but Manufacturing resources for only one computer.

Order queuing defines the sequence in which incoming Request Orders are serviced. When multiple Request Orders are waiting for service, they can be queued according to a variety of policies, including:

- **First Come, First Served (FCFS)**. This is the fairest policy, but one that does not allow one to give priority to customers that require a higher level of serviceability.
- **Priority.** Orders with highest priority are serviced first. Priority is a property that is assigned to each order when it is created.
- **Due Date**. Orders with the earliest due date are serviced first.

In addition to these policies for sequencing orders, some assembly operations may take place according to specified production schedules, often in batches. Such schedules are derived from the Supply Plans received from Supply Planning functions.

Product allocation defines the policy used to assign finished goods inventory Request Orders. For example:

- Under **Greedy Allocation**, when a Request Order is received, the Manufacturing function checks its finished goods inventory. If inventory is available to completely fill the order, the process allocates finished goods to the order. If there is insufficient inventory, the next step depends upon whether the order accepts partial shipments. If it does, the process can allocate a partial amount of inventory to the order and put the remaining order on queue. If it does not, the complete order is put on queue until inventory can be allocated to completely fill the order.
- Under **Periodic Allocation**, all Request Orders are put on queue when they arrive. Periodically, based on a specified review period, the order at the head of the queue is examined to see if finished goods inventory is available. If inventory is available to completely fill the order, inventory is allocated to the order and the next order on the queue is examined. If inventory is not available, the next step depends upon whether the order accepts partial shipments. If it does, the process can allocate a partial amount of inventory to the order and put the remaining order back on queue. If it does not, the complete order is put back on queue until inventory can be allocated to completely fill the order.
- Under **Reserved Allocation**, another supply chain function must reserve inventory at this Manufacturing function prior to sending a Request Order.

Inventory replenishment must maintain enough inventory to satisfy customer demand while controlling inventory costs. A Manufacturing function maintains inventory in logical storage areas called *buffers*. Two types of buffers are maintained, *input buffers* for raw materials and *output buffers* for finished goods. A Manufacturing function sends out Request Orders to restock its input buffers. The assembly process transforms raw materials in input buffers to finished goods in output buffers.

Raw materials can be either outsourced or insourced. Outsourced raw materials are ordered from another supply chain function. Insourced raw materials are manufactured at the same function where they are used.

A specified inventory replenishment policy determines when and how a Manufacturing function generates Request Orders to restock its buffers. Here are some examples of replenishment policies:

- **Continuous Replenishment.** The buffer is restocked whenever the inventory level in the buffer falls below a specified reorder point.
- **Periodic Replenishment.** The buffer is restocked periodically based on a specified review period, but only if the inventory level in the buffer is below its reorder point.
- **Build-To-Order (BTO).** This policy maintains minimum inventory by restocking a buffer only if a Request Order arrives and inventory is not available to fill the order.
- **Build-To-Plan (BTP).** Buffers are restocked according to Supply Plans received from Supply Planning functions.

In a Manufacturing function there are two types of material handling, *inbound handling* (dock-to-stock) and *outbound handling* (stock-to-dock). A Manufacturing process must model both the time and cost of material handling. Material handling cost can be modeled in a number of ways, for example:

- Cost per order
- Cost per unit of weight
- Cost per unit of volume
- Cost per order per hour
- Cost per unit of weight per hour
- Cost per unit of volume per hour

Partial pallets are usually much more costly to handle than full pallets – this must also be captured in the cost model.

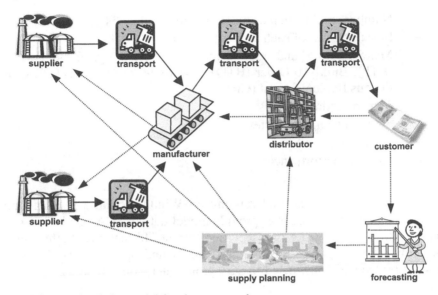

Fig. 2.3. Supply chain model for the case study

2.3 Strategic Uses of Supply Chain Simulation

Researchers at IBM have been active in supply chain simulation for many years. The bulk of our work in this area has been strategic in nature – standalone, one-time simulations used to make structural or policy decisions in IBM's internal supply chain or a supply chain of an IBM customer. For example, during the 1990's IBM reengineered its global supply chain to achieve quick responsiveness to its customers with minimal inventory. To support this effort, we developed a supply chain analysis tool called the Asset Management Tool (AMT). AMT integrated graphical process modeling, analytical performance optimization, simulation and activity-based costing into a system that supports quantitative analysis of extended supply chains. IBM used AMT to study such issues as inventory budgets, turnover objectives, customer-service targets and new product introductions. It was used at a number of IBM business units and their channel partners. AMT benefits included over $750 million in material costs and price-protection expenses saved in 1998. IBM was awarded the prestigious Franz Edelman award from INFORMS in 1999 for this work (Lin et al. 2000). AMT was later made into an IBM product called the Supply Chain Analyzer (SCA) which was used in consulting engagements by IBM Global Services (Bagchi et al. 1998). SCA was used to perform strategic studies for IBM customers addressing issues which include:

- Number and location of manufacturers and DC's
- Stocking level of each product at each site
- Manufacturing and replenishment policies, e.g. Build-To-Plan (BTP), Build-To-Order (BTO), Assemble-To-Order (ATO), Continuous Replenishment (CR)
- Transportation policies
- Supply planning policies
- Lead times
- Supplier performance
- Demand variability

SCA was a standalone tool running on Windows with a user-friendly graphical interface. In order to provide model data to SCA one had to prepare a number of flat files in a specified format. In many cases this was a one-time manual process using query tools and spreadsheets. In some cases a bridge was constructed to SCA from enterprise databases.

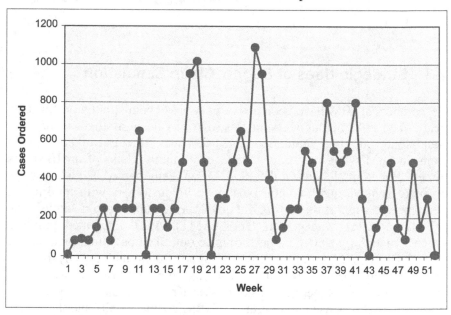

Fig. 2.4. Simulated customer demand

2.3.1   Case Study: Build-to-Plan vs. Continuous Replenishment

For this case study, we constructed a model that has some of the intricacies of a real supply chain, but is much simpler so that it can be easily explained. The model is shown in Figure 2.3. In this model, the *Customer* process generates daily orders for just one product type. The simulated customer demand has seasonality and promotional spikes, as shown in Fig. 2.4.

Customer demand is fulfilled by a distribution center, as shown in Fig. 2.3. There is a one day stock-to-dock delay at the distribution center and a one day transportation delay from the distribution center to the customer. Customer orders must be satisfied in three days or less to be considered on time.

The distribution center is stocked by a manufacturing plant, as shown in Fig. 2.3. The distribution center and the manufacturing plant are owned by the same company, MFGCO. There is a one day stock-to-dock delay at the manufacturing plant, a one day transportation delay to the distribution center and a one day dock-to-stock delay at the distribution center, totaling a three day lead time from manufacturing to the distribution center. Note that customer orders cannot be delivered on time if the distribution center has to backorder the stock from manufacturing, because the total lead time to the customer would be five days and customers require delivery in three days or less. Therefore, the distribution center must keep enough finished goods on hand to satisfy its daily demand.

The manufacturing plant assembles products using components obtained from two suppliers, as shown in Figure 2.3. The lead time from each supplier is one day. The suppliers are not owned by MFGCO, but transportation from the suppliers is paid for by MFGCO. MFGCO also pays for transportation from the manufacturing plant to the distribution center. Transportation from the distribution center to the customer is paid for by the customer.

Three costs are calculated when this model is simulated:

- Inventory holding costs – 15% of product cost
- Material handling costs – at representative per-pound rates
- Transportation costs – at representative less-than-truckload rates

To illustrate a typical simulation exercise, we created an AS-IS case in which the network is supplied by Build-To-Plan (BTP) logic. We then created a TO-BE case in which the network is instead supplied by Continuous Replenishment and we compared the resulting costs to the AS-IS case.

For the AS-IS case, a customer forecast was created which has a 20% error margin relative to the actual demand. The forecast is maintained in weekly buckets. Once every four weeks, an MRP system explodes the customer forecast into a build plan for the manufacturer and suppliers and a replenishment plan for the distribution center. This manufacturing policy is commonly referred to as Build-To-Plan (BTP). To simulate this, we embedded an IBM-developed MRP tool (Dietrich et al. 1995) within the simulator, labeled as *Supply Planning* in Figure 2.3. The build plans and replenishment plans are generated in weekly buckets. Note that in order for the distribution center to have enough on-hand stock to satisfy demand at the beginning of each week, manufacturing must build the forecasted demand the week before.

For the TO-BE case, the same customer forecast is used, but the inventory at each stocking location is reviewed daily. The forecast is used by an optimization program (Ettl et al. 2000) to set the reorder point at each stocking location each day. Whenever a reorder point exceeds the on-hand inventory, a replenishment order is sent out to make up the difference. This manufacturing policy is commonly referred to as Continuous Replenishment (CR).

The goal of both simulations was to service customer orders at or near 100%, based on the three day requirement. Figure 2.5 shows the simulated inventory levels at the Distribution Center for the AS-IS case. Fig. 2.6 shows the simulated inventory levels at the Distribution Center for the TO-BE case. Note that both figures represent simulation outputs. In both figures, the ragged lines represent the actual inventory and the smooth lines represent the average inventory. Note the similarity of these inventory profiles to the customer demand shown in Figure 2.4. As you can see, the AS-IS case (BTP) must keep more inventory on hand than the TO-BE case (CR) because the AS-IS case builds a week ahead of time while the TO-BE case reviews the inventory levels each day.

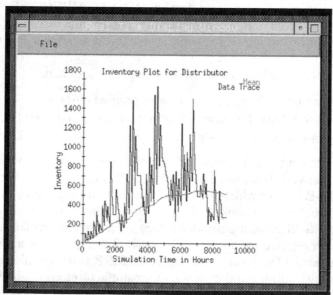

Fig. 2.5. Simulated inventory levels at the Distribution Center for the AS-IS case

Figure 2.7 summarizes the results of the simulation runs. The shipments are the same in both runs, as are the revenue and the transportation costs to the customer. The transportation and handling costs at MFGCO are slightly lower in the TO-BE case, but that is primarily due to slight differences in the initial inventory of the two simulation runs. The only major difference is in the inventory, which is reduced by 62% in the TO-BE case.

Continuous Replenishment has many other benefits. For one thing, expired products usually decrease when Continuous Replenishment is used. These savings can be calculated by a supply chain simulator, but in this simple example, there were no expirations because both cases were fairly lean. In addition, the daily review that is associated with Continuous Replenishment can lead to increased full pallet orders and direct manufacturing shipments. These savings can also be calculated by a supply chain simulator.

The simple example presented here shows only a small fraction of the features and power of a supply chain simulator. Consultants and process engineers often wonder how to convince upper management to convert from an outdated process to a modern process (e.g. Build-To-Plan to Continuous Replenishment). Simulation is one way to do this. When an AS-IS simulation is properly validated against historical data, upper management can have confidence that it represents their company. Then, when a TO-BE policy is tested by the simulation, there is quantitative evidence to support or argue against making the change.

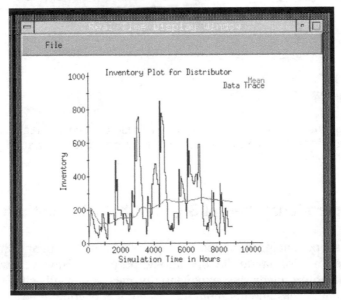

Fig. 2.6. Simulated inventory levels at the Distribution Center for the TO-BE case

Measurement	AS-IS	TO-BE	Savings	Savings%
Shipments (cases)	18,277	18,283		
Revenue	$548,310	$548,490		
Customer Transportation	$3,655	$3,657		
MFGCO Transportation	$6,771	$6,535	$236	3
MFGCO Handling Cost	$14,137	$13,826	$311	2
MFGCO Inventory Cost	$38,751	$14,864		
MFGCO Carrying Cost @ 15%	$5,813	$2,230	$3,583	62

Fig. 2.7. Simulation results

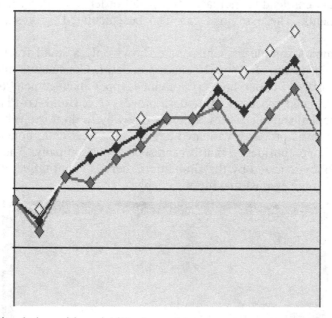

Fig. 2.8. Simulation with variability is used to generate a control region for inventory. The middle line represents expected inventory. The upper and lower lines are control bounds

2.4   Operational Uses of Supply Chain Simulation

In IBM, supply chain simulation tools have been used primarily for strategic studies, not for day-to-day operational purposes. However, with computational power increasing exponentially each year and with enterprise data becoming substantially more integrated, it is becoming possible to use this technology for operational decisions. Slowly, supply chain simulation is spreading into the weekly and daily operations of enterprises. In the future, this transition will be made easier by the following advances:

- Simulation speed is increasing due to drastic improvements in computer technology coupled with careful design of simulation granularity.
- Simulation model data will become more integrated with enterprise data. As Business Activity Monitoring (BAM) (April and Margulius 2002) and Business Performance Management (BPM) (www.ibm.com/ software/info/ topic/perform/resources.html) grow in popularity, simulation data will be more readily available in data warehouses, reducing the startup cost to create a simulation model.
- What-if simulation of alternatives will increasingly become part of decision-making processes.
- Business users of simulation technology will be presented with customized screen flows, not general-purpose simulation tooling. They may not even know that they are using a simulation tool.
- Simulation tools will be web-enabled. Business process management is shifting to the web and data is readily available on the Internet. Modern web portal technology supports customizable user interfaces.

The following scenarios illustrate the operational use of supply chain simulation:

- **Process control**. Simulation is used to predict the metrics of a process in an upcoming time period. The process is then tracked against the simulated results. For example, an IBM division uses simulation to predict their future product inventory levels at the beginning of each quarter (see Figure 2.8). Based on various uncertainties specified in the simulation, it is possible to generate lower and upper bounds for statistical control purposes. Actual inventory is then tracked against the control limits for early detection of unexpected situations.
- **Decision support**. When unexpected situations arise, there are often a number of alternative actions that can be taken. Simulation can be used to assess the benefit and risk of each potential response (Lin et al. 2002). For example, the potential responses to a late supplier delivery may include setting higher inventory targets, using a different supplier, and doing nothing. These alternatives can be simulated under the current business conditions to predict the profit, cost and serviceability for each alternative. From multiple runs these predictions can summarized in stochastic terms to estimate risk.

- **Trend detection**. Simulation can be used to predict potentially harmful business trends using information sensed from the business environment (Lin et al. 2002). Figure 2.9 shows an example in which simulation is used to predict customer demand.

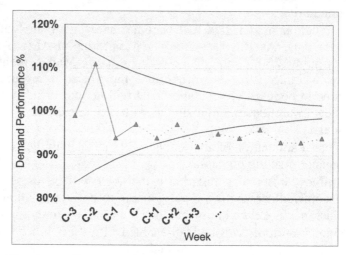

Fig. 2.9. The dotted line shows the customer demand, simulated under current business conditions to see if it stays within established control limits

References

April, C., Margulius, D. (2002) BAM to speed app reports, *InfoWorld* 24 (44): 27–28

Bagchi, S., Buckley, S., Ettl, M., Lin, G.Y. (1998) Experience using the IBM Supply Chain Simulator, *Proceedings of the 1998 Winter Simulation Conference*, Washington, DC, pp 1387–1394

Dietrich, B., Connors, D., Ervolina, T., Fasano, J.P.,, Lin, G., Srinivasan, R., Wittrock, R., Jayaraman, R. (1995) Production and Procurement Planning Under Resource Availability Constraints and Demand Variability, *IBM Research Report RC-19948*, Yorktown Heights, NY

Ermakov, S., Melas, V. (1995) *Design and Analysis of Simulation Experiments*. Kluwer Academic Publishers, Dordrecht, Netherlands

Ettl, M., Feigin, G., Lin, G., Yao, D. (2000) A Supply Network Model with Base-Stock Control and Service Requirements. *Operations Research* 48 (2): 216–232

Feigin, G., An, C., Connors, D., Crawford, I. (1996) Shape Up, Ship Out. *ORMS Today* 23 (2)

Katircioglu, K., Grey, W., Shi, D., Bagchi, S. (2002) Impact Analysis of e-Business Metrics and Policies. *INFORMS Annual Meeting*, San Jose, CA

Lin, G., Ettl, M., Buckley, S., Bagchi, S., Yao, D.D., Naccarato, B.L., Allan, R.,
    Kim, K., Koenig, L. (2000) Extended-Enterprise Supply-Chain Man-
    agement at IBM Personal Systems Group and Other Divisions, *Inter-
    faces* 30 (1): 7–21

Lin, G., Buckley, S., Cao, H., Caswell, N., Ettl, M., Kapoor, S., Koenig, L., Katir-
    cioglu, K., Nigam, A., Ramachandran, B., Wang, K. (2002) The
    Sense-and-Respond Enterprise. *OR/MS Today* 29 (2): 34–39

Supply-Chain Council (2000) Supply-Chain Operations Reference-Model: Over-
    view of SCOR Version 5.0. Pittsburgh, PA: Supply-Chain Council
    Inc.

3 Inventory Management in High Technology Value Chains

Feng Cheng, Markus Ettl, Grace Lin, and David D. Yao

3.1 Introduction and Overview

In the computer industry we see dramatic reductions in the price of computers, driven by advances in technology and competitive forces pushing towards lower margins. Product life cycles are collapsing to months rather than years, giving companies less time to recover product development costs and increasing the pressure to rapidly and flawlessly commercialize new technology. We are constantly being challenged to improve on productivity throughout the enterprise. We continue reengineering business processes and eliminating steps, and over the past several years our attention has increasingly turned toward managing the value chain.

IBM and its partners and competitors seek operational and financial performance improvement through reduced product development cycles, operational efficiency, and better customer responsiveness. This quest for excellence has become significantly more complex as enterprises no longer compete as stand-alone entities. The success is achieved through value chain optimization and collaboration among all value chain participants, from OEM's, Tier-1's, and lower tier suppliers to distributors, trading partners, and retailers. Deploying common business processes across distinct operating entities allows participants to share decision-making, workflows, and capabilities in pursuit of lower costs and greater efficiency.

A significant challenge, and opportunity, for IBM is that we have one of the most vertically integrated supply chains in the industry. We manufacture most major assemblies in our computers. As a result, IBM's integrated supply chain is even more complex and difficult to manage. Our supply chain is under constant pressure to move towards the assembly of components from a vast array of outside suppliers, and to sell components we make to the marketplace. More and more, we deliver components to internal business units and also sell them to competitors, some who manage their supply chains and assembly operations to deliver higher levels of profitability than IBM derives from the same markets. Clearly, such an environment makes managing the extended supply chain critical to our success.

The business environment in the electronics industry, which is characterized by volatility and velocity, requires tools and applications that can recommend timely supply planning decisions that optimize profits and balance business risk. Standard enterprise applications such as enterprise resource planning (ERP), customer relationship management (CRM), and supply chain management (SCM) systems are effective in managing hundreds of product and service offerings, but they often lack in high-quality decision making.

In this chapter, we describe analytical models and tools that we have developed to support IBM business units in their effort to manage inventory and improve value chain operations. In IBM's businesses, inventory-driven costs, which include price protection, financing, inventory write-downs (price erosion), and inventory write-offs (obsolescence) are tremendous cost drivers outweighing all others in terms of impact on business performance. The complexity of the end-to-end value chain makes it difficult to determine where to hold safety stock to minimize inventory costs, and provide a committed level of service to the final customer. We developed analytical optimization models for finding the optimal placement of safety stocks in multi-echelon value chains that are subject to forecast, lead time, and attach-rate uncertainty. We will describe the successful application of these analytical models by cross-functional teams within IBM, our suppliers, and our customers. We also discuss how the teams have used the models to allocate component inventories, reduce finished goods inventories, manage product variety, and improve forecast accuracy.

To put our discussions in context, we shall focus on three specific value chain architectures – the complex configured hardware value chain, configure-to-order value chain, and semiconductor value chain. Our model for the complex configured hardware value chain emphasizes the optimization of products with multi-tier complex bill of material. The main objective is to minimize the total inventory cost while meeting the target customer service levels. This type of value chains are normally subject to long assembly or testing lead times, high forecast inaccuracy and skewed demand. The configure-to-order value chain has a simple, one step assembly process with a very short assembly time. Its main challenges are low profit margin, short product life cycle, and uncertain supplier lead times, while requiring quick responsiveness, low cost, and a high level of product variety. The optimization model we shall present emphasizes the optimal control of the component inventory to ensure low costs while meeting customer serviceability. Our model for the semiconductor value chain involves an arborescent process-product structure and location-based budget constraints, with the objective to minimize overall delinquency costs (penalties for unmet demand).

The rest of the chapter is organized as follows. We start with an overview in section 3.2 of different value chain architectures in high technology industries, focusing on the three types highlighted above. Details of the modeling and applications are elaborated in the next three sections. In section 3.3, we introduce a multi-echelon inventory model for complex configured products, and discuss its application in IBM's hard disk drive supply chain. In section 3.4, we develop an optimization algorithm for safety stock placement in a configure-to-order supply chain with high-volume, high variety products, and describe its application at IBM's personal computer division. In section 3.5, we describe an inventory model for a semiconductor supply chain, and present our application experience at a large US semiconductor manufacturer. Non-technical readers may skip the details of the optimization modeling in the first part of sections 3.3 to 3.5 and proceed directly to the case studies. We conclude the chapter with a summary in section 3.6.

A very brief note on related readings: More details of the application sections can be found in Ettl and al. (2000) and Lin et al. (2000) for section 3.3, Cheng et al. (2002) for section 3.4, and Brown et al. (2001) for section 3.5. The three edited volumes, De Kok and Graves (2003), Song and Yao (2001) and Tayur et al. (1999) collect many recent research works on various aspects of supply chain management. Background materials in inventory theory can be found in Zipkin (2000).

3.2   High-Technology Value Chains

In this section, we describe three examples of value chain applications in high-technology industries. For each application, we identify end-to-end management processes that enable their business objectives.

- **Complex Configured Hardware Value Chains.** Standard part number based product offerings with complex bills of materials. Fulfilling customer orders through vendor-managed inventory hubs outside of the customer's manufacturing plant.
- **Configure-to-Order Value Chains.** Customizing products and solutions quickly to customer requirements. Developing configure-to-order capabilities for direct selling through the Internet.

- **Semiconductor Value Chains.** Implementing postponement strategies to defer customer-specific configurations until as late as possible. Determining the right safety stock policies at the inventory postponement points.

3.2.1 Complex Configured Hardware Value Chains

Complex configured hardware value chains support part number based components and products with complex bills-of-materials. Customer orders are typically submitted through enterprise web sites for large enterprise customers, business-to-business portals for business partners, and public websites for consumers and small and medium business customers. Advanced Planning System (APS) applications reconcile customer forecasts with existing supply, and send allocations to a fulfillment system. The fulfillment system schedules orders, calculates estimated customer arrival dates, and sends manufacturing orders to a floor control system to manage the assembly of products.

The manufacturing process of hard disk drives (HDDs) is a typical example of a complex configured hardware value chain. IBM's Storage Systems Division, now Hitachi Global Technologies, produces disk drives for the OEM market, as well as supplies internal IBM business units. The supply chain for HDDs is complex due both to the vertically integrated nature of IBM and the customers' requirement for Just-In-Time (JIT) inventory hubs, where IBM inventory is stocked outside of a customer's factory.

HDDs are sold to OEM customers, distributors, and used in other IBM products such as personal computers and servers. Two of three product segments, mobile and performance HDDs, are manufactured in Thailand and Hungary, and server HDDs are manufactured in Singapore. In addition to the JIT hubs, the distribution network includes Logistics Centers and internal IBM direct shipments as illustrated in Figure 3.1. When HDDs are in excess supply at one hub but needed at another customer's hub, they can be reconfigured at a Logistics Center and then redirected. While reconfiguration occurs infrequently, it is fairly common to need to reship an HDD for a given customer from one hub to another. HDDs are assembled from non-configured "vanilla" drives (called ISOs) at the manufacturing plants, and shipped to a customer JIT hub fully configured for that customer. When the JIT hub inventory drops below a reorder point, a pull signal is sent to the plant to configure the ISO's for that hub. The pull strategy requires forecasting to determine safety stock levels that protect against variations in demand.

Because of limited production capacity, the plants typically overproduce (versus demand) early in the quarter to meet peak demands towards the end of the quarter. This demand skew is much more prevalent in Logistics Centers than it is for JIT hubs. Because the configuration process is relatively simple, and because demand for the vanilla drives is easier to predict than HDD demand, it is beneficial to delay customization of an ISO as long as possible.

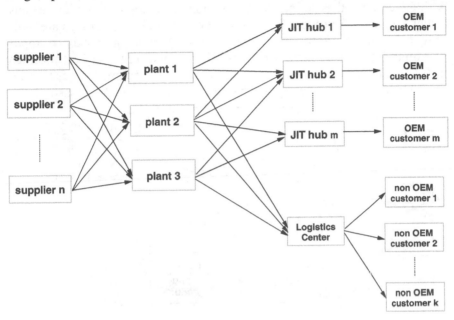

Fig. 3.1. Example of a Complex-Configured Hardware Value Chain

3.2.2 Configure-To-Order Value Chains

In computer assembly value chains, products are offered as either fixed configurations or open configurations. Fixed configurations have an individual material code that is referred to in customer orders. Open configurations, in contrast, can be fully configured by the customer. Here, the customer can navigate from a brand, series, or family via a web-based configurator, which shows the options compatible with the selected machine. The configurator lets the customer choose from a selection of processors, hard disk drives, network access cards, graphics cards, and memory sizes. Only the base unit and standard components (called building blocks) have individual material codes. Whereas fixed configurations are normally made-to-stock, open configurations are assembled-to-order after the customer order is received (configure-to-order, CTO). Figure 3.2 illustrates a configure-to-order supply chain.

The purpose of forecasting in computer assembly value chains is to produce an accurate component forecast. For fixed configurations, the bills-of-materials of the fixed configurations are simply exploded to component level. For open configurations, component forecasts are derived from a collaborative forecast for a customer segment (or product family), together with attach-rates that define the distribution of components within the customer segment (or product family). As the component lead times are much longer than the assembly lead times, the component forecast is an important input into procurement decisions. The focus of inventory management in CTO value chains is shifting from configured machines towards building blocks, which are replenished based on the component forecast.

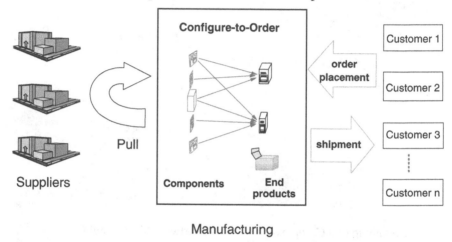

Fig. 3.2. Example of a complex configured hardware value chain

The configure-to-order paradigm has been widely accepted in the electronics industry. Open configurations offer higher product variety, and hence often results in broader market coverage and increased demand volume. Postponing the final assembly in configure-to-order operations provides flexibility in terms of product variety, and achieves resource pooling in terms of maximizing the usage of component inventory. CTO is an ideal business model for mass customization, and provides quick response time to order fulfillment.

3.2.3 Semiconductor Value Chains

The semiconductor industry has one of the most complex manufacturing processes and value chains. Equipment necessary for production is extremely capital expensive and difficult to install. The semiconductor industry is at the beginning of a complex network of value chains, which often leads to tremendous demand fluctuations and high uncertainties of demand forecasts. As a result, many semiconductor companies have outsourced key value chain activities to electronic manufacturing service providers (EMS) to focus on core competencies, such as product design, inventory management, order fulfillment, and utilizing production capacity efficiently.

The production process of semiconductors consists of two stages called front-end and back-end that are separated by a die bank. The front-end consists of wafer fabrication and wafer testing, whereas the back-end performs die bonding, assembly, and module testing. The production process is often split over several locations in different geographies, e.g. front-end facilities in South-East Asia, and back-end facilities in Europe and North America. The production lead time of the front-end is eight to twelve weeks. After testing, wafers are sliced into individual chips that are subsequently used to produce different finished goods. The tested chips are stored in a die bank. The back-end operates in a make-to-stock or make-to-order mode. In the assembly stage, chips supplied from the die bank are connected to a platform, bonded, and sealed in plastic. The finished modules are shipped to a customer or stored in finished goods inventory. The lead time of the back-end is four to six weeks. Figure 3.3 illustrates a semiconductor value chain.

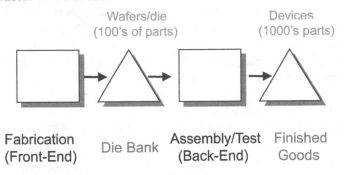

Fig. 3.3. Example of a semiconductor value chain

The customers of semiconductor manufacturers are usually large resellers or OEM's that can have considerable influence. In this industry, high level of customer serviceability is critical since although an integrated circuit may only be one component of hundreds in a customer's product, a late delivery may shutdown a production line.

3.3 A Multi-Echelon Inventory Model for Complex Configured Hardware

A key driver to achieve supply chain optimization is to manage and reduce uncertainty, and to maintain flexibility so as to adapt to market changes quickly. How much inventory budget is needed to achieve good customer serviceability? How much inventory do we need at which locations to meet required service levels? What is the effect of changes in supplier lead times, or supplier practices? What is the effect on required inventory levels, of changes to where and how we assemble products? IBM's Asset Management Tool (AMT) was designed to answer these questions.

3.3.1 AMT: The Optimization Engine

The optimization engine performs AMT's main function: quantifying the trade-offs between customer service levels and the inventory in the supply network. The objective is to determine the safety stock for each product at each location in the supply chain to minimize total inventory investment. Below we present an overview of this optimization model, referring the complete technical details to Ettl et al. (2000).

The supply chain is modeled as a multi-echelon network. Each stocking location in the network is treated as a queueing system that incorporates an inventory control policy: the base-stock control, with the base-stock levels being decision variables. To analyze such a network, we develop an approach based on decomposition. The key idea is to analyze each stocking location in the network individually, and to capture the interactions between different stocking locations through their *actual lead times*.

We model each stocking location by a queue with batch Poisson arrivals, and infinite servers with service times following general distributions, and this model is denoted as $M^Q / G / \infty$ in queueing notation. To do so, we need to first specify the arrival and the service processes. The arrival process at each location is obtained by applying the demand explosion technique in standard MRP (materials requirement planning) to the product structure. The batch Poisson arrival process has three main parameters: the arrival rate, and the mean and the variance of the batch size. Therefore, it allows us to handle many forms of demand data using a three-parameter fit. For instance, demand in a certain period can be characterized by its min, max and the most likely value. The service time is the actual lead time at each stocking location. Figure 3.4 illustrates how the actual lead time, $\widetilde{L}_i$ is calculated from the nominal lead time (e.g., production or transportation time), L_i, along with the fill rate, f_i, of location i's supplier j. In particular, when the supplier has a stockout, which happens with probability $1 - f_i$, the actual lead time at i has an additional delay of τ_j, which is the time required for j to produce the next unit to supply i's order. In our model, the estimation of τ_j is derived from a Markov chain analysis.

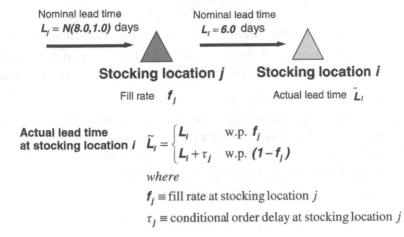

$$\tilde{L}_i = \begin{cases} L_i & \text{w.p. } f_j \\ L_i + \tau_j & \text{w.p. } (1-f_j) \end{cases}$$

where

$f_j \equiv$ fill rate at stocking location j

$\tau_j \equiv$ conditional order delay at stocking location j

Fig. 3.4. Actual lead times

With the arrival and service processes in place, we can analyze the queue and derive performance measures such as inventory, backorder, fill rates, and customer service levels. The number of jobs in the $M^Q / G / \infty$ queue i, X_i, is the key quantity in our analysis. The on-hand inventory I_i and the backorder level B_i relate to X_i through the following simple formulas:

$$I_i = [R_i - X_i]^+, \text{ and } B_i = [X_i - R_i]^+, \tag{3.1}$$

where R_i is the base-stock level, and $[x]^+ := \max\{x, 0\}$.

Through the equations in (3.1), the distributions of both I_i and B_i can be related to the distribution of X_i. To alleviate the computational burden in large-scale applications, we choose to derive the mean and the variance of X_i, and approximate it with a normal distribution. Both the mean and the standard deviation of X_i, denoted μ_i and σ_i, depend on the actual lead time (explained above, and used as the service time in the queue model), and can be derived from queueing analysis (refer to Ettl et al. 2000). Then, we write:

$$X_i = \mu_i + \sigma_i Z, \tag{3.2}$$

where Z denotes the standard normal variate. Similarly, we can express the base-stock level as follows:

$$R_i = \mu_i + k_i \sigma_i, \tag{3.3}$$

where k_i is the so-called safety factor. This way, we turn the decision variables from the base-stock levels to the safety factors. Also note that the base-stock level consists of two parts: μ_i, the work-in-process, or pipeline inventory; and $k_i \sigma_i$, the safety stock.

The objective of the optimization model is to minimize the total inventory capital. At each stocking location, there are two types of inventory: the finished goods, or on-hand inventory, and the pipeline inventory. The expected pipeline inventory is simply μ_i; the expected on-hand inventory follows from combining (3.2) and (3.3) with (3.1):

$$\mathbf{E}(I_i) = \mathbf{E}[R_i - X_i]^+ = \sigma_i \mathbf{E}[k_i - Z]^+.$$

Let

$$H(x) := \mathbf{E}[x - Z]^+ = \int_{-\infty}^{x} (x - z)\phi(z)dz = x\Phi(x) + \phi(x) \tag{3.4}$$

where ϕ and Φ denoting, respectively, the density function and the distribution function of Z. Then,

$$\mathbf{E}(I_i) = \sigma_i H(k_i). \tag{3.5}$$

Hence, the objective of our optimization model is

$$\min_{\mathbf{k}} C(\mathbf{k}) := \sum_{i \in S} \mathbf{E}[\hat{c}_i \mu_i + c_i \sigma_i H(k_i)]. \tag{3.6}$$

Here S is the set of all stores; c_i is the unit cost of finished goods inventory at store i; and $\hat{c}_i$ is the unit cost of pipeline inventory,

$$\hat{c}_i := \tfrac{1}{2}(c_i + \sum_{j \in S_{>i}} c_j u_{ji}),$$

where $\mathbf{S}_{>i}$ is the set of upstream stores of i, and u_{ji} is the usage count, the number of units in store j that is needed to make each unit in store i.

The constraints of the optimization model are the required customer service levels. They are represented as the probability (e.g., 95% or 99%) that customer orders are filled within a given due date. We first derive the required fill rate for each end product so as to meet the required customer service level. This fill rate relates to the actual lead times of all upstream stocking locations, via the BOM structure of the network. Therefore, our model captures the interdependence at different stocking locations, in particular the effect of base-stock levels and fill rates at each stocking location on the service level of the end product.

To allow fast execution of the optimization, we derive analytical gradient estimates in closed forms. Consider stocking location j, or the j-th term in the objective function. First, we derive the partial derivative w.r.t. k_i (the safety factor) for stocking location j. For each immediate downstream stocking location, the partial derivative involves the mean and variance of the number of arrivals over the service time (i.e., the actual lead time) in the queue model. For stocking locations further downstream, we simply ignore their derivatives, since their actual lead times will be weighted by multiples of no-fill rates, which become negligible as the stocking locations become farther downstream.

This way, we have a constrained non-linear optimization model, with the gradients explicitly derived. A conjugate gradient search procedure is used to generate the optimal solution. As the objective function has a quite rugged surface, we improve upon local minima by following several heuristic procedures. For instance, evaluating the objective function at a reasonably large number of randomly generated points, and selecting the best point to start the gradient search.

3.3.2 Case Study: IBM's Hard Disk Drive Value Chain

In late 1997, many of IBM's HDD customers had adopted the practice of vendor-managed inventory, which required IBM to establish Just-In-Time (JIT) inventory hubs nearby the customer's plants. The inventory in the JIT hubs was managed and owned by IBM until it was pulled from the hub by the customer.

As more and more JIT hubs were established and an increasing proportion of HDD inventory flowed through the hubs, it became more important to minimize the amount of product in the hubs. However, it was unclear what the impact of delayed customization would be on the supply chain and, in particular, on inventory levels and customer serviceability. It was also unclear what levels of safety stock, and therefore what reorder points were needed in the JIT hubs to fulfill customer service requirements. IBM's Corporate Headquarters formed a cross-functional team whose charge was to assess the impact of delayed customization on the HDD supply chain, and to recommend optimal inventory levels for the JIT hubs.

The primary objective of the team was to use AMT to determine optimal base-stock levels (or reorder points) at the JIT hubs, to implement a con-figure-to-pull strategy at these hubs, and to improve inventory turnover while meeting customer service requirements. AMT allowed the explicit representation of assembly and transport operations. Bills-of-materials and manufacturing lead times (if outsourced, replenishment lead times) were extracted from SAP and incorporated into the model. Forecast errors were estimated at the product family level to provide demand distribution data for our analysis.

We created end-to-end simulation models for each product family, and validated these models against historical data to assure that the information and materials flow through the HDD value chain was captured accurately. Comparing key output performance measures of our model (i.e., finished goods inventory at JIT hubs and plants) with historical actuals, we found that JIT hub inventory matched at a part number level to within 8% of ac-tual on average. JIT hub inventory matched at a product family level to within just 6% of actual, and plant finished goods inventory matched at a product family level to within 8% of actual. On-time serviceability matched at close to 100% at part number and order level. An example of the validations is shown in Figure 3.5.

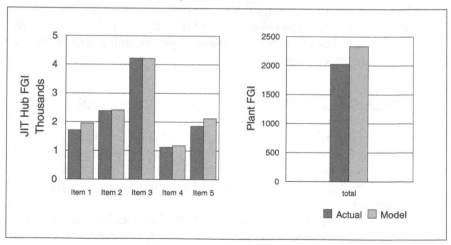

Fig. 3.5. Comparison of FGI between simulation model and historical actuals

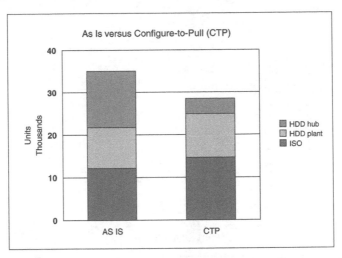

Fig. 3.6. Comparison between build-to-stock (As-Is) and configure-to-pull (CTP)

Subsequently, we constructed simulation models for a select set of high-volume HDD products based on historical inventory and demand data. One model represented the current business process, whereas the other represented a "configure-to-pull" business process with optimized inventory buffers at the JIT hubs. The comparison between the two models showed significant savings in hub inventory, averaging about 69%, can be achieved with configure-to-pull without sacrificing serviceability. The savings were achieved with only a 21% increase in plant inventory of non-configured vanilla drives (ISOs) as shown in Figure 3.6.

By varying the time needed to customize an ISO in the plant, we then established the relative impact of reducing the manufacturing lead time on inventory. We found that reducing the lead time by one day, on average, leads to 8% savings in total inventory. These results were fairly consistent across product families. Figure 3.7 depicts this sensitivity.

In summary, the study showed that delayed customization can significantly reduce finished goods inventory at the JIT hubs, and reduce costs for reshipping products without sacrificing serviceability (i.e., without increasing stockouts at the hubs). It also showed that moving towards consumptive pull and shortening the manufacturing and configuration lead times further reduces costs in the form of inventory reduction, which was a key step to solidify management support for eventual implementation of the configure-to-pull process design change.

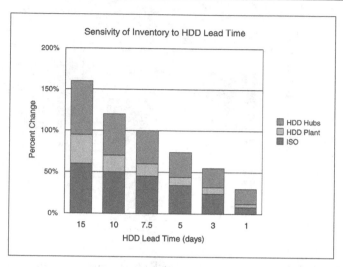

Fig. 3.7. Sensitivity analysis results on reducing manufacturing lead times

3.4   Configure-to-Order for High-Volume High-Variety Products

The focus of the following study is on the inventory-service tradeoff in configure-to-order (CTO) manufacturing in IBM's personal computer value chain. It is part of a larger project that aimed at helping IBM's personal computer division to migrate from fixed configurations to a configure-to-order operation where customer orders are taken from the Internet. We have developed and applied an analytical trade-off model to evaluate three scenarios: assess the cost/benefit of a building-block based manufacturing operation; compare the forecasting of fixed configurations versus customer segment and attach-rate forecasting; and assess the effect of increasing product variety on inventory.

3.4.1   The Optimization Model

We consider a hybrid model, by which each end product is assembled to order from a set of components, which, in turn, are built to stock. In other words, no finished goods inventory is kept for any end product, whereas each component has its own inventory, replenished from a supplier following a base-stock policy.

Each component inventory is indexed by i, $i \in S$ where S denotes the set of all components. Associated with each component is a "store", where the inventory is kept. In the configure-to-order (CTO) environment, there is no pre-specified product "menu"; in principle, every order can require a distinct set of components. Let $\mathbf{M}$ denote the set of product families (or market segments) that order the same set of components. For instance, $\mathbf{M} = \{$ low-end machines, high-end machines, servers $\}$; or $\mathbf{M} = \{$ consumers, small businesses, corporations $\}$.

Let $D_m(t)$ denote the demand associated with product family m in period t. Each order of type m requires a random number of units from component i, denoted as X_{mi} which takes on non-negative integer values. Denote:

$$\mathbf{S}_m := \mathbf{S} - \{i : X_{mi} \equiv 0\} \text{ and } \mathbf{M}_i := \mathbf{M} - \{m : X_{mi} \equiv 0\}.$$

That is, $\mathbf{S}_m$ denotes the set of components used in type m products, whereas $\mathbf{M}_i$ denotes all the product families that use component i. Here, $X_{mi} \equiv 0$ means $P(X_{mi} = 0) = 1$.

The first step in our analysis is to translate the end-product demand into demand for each component. This is done through the bill-of-material structure for the products. (Alternatively, component demand can be derived through forecast data on aggregated demand over market segments and the attach-rates of the components. For example, 90% of products sold to large corporations, and 50% in the small business and consumer segments will use high-end processors, and so forth.)

There are two kinds of lead times: those associated with the components (inbound) – the time for the supplier of component i to replenish to store i once an order is placed; and those associated with the end products (outbound) – including the time to process orders, the assembly/reconfiguration time, and the transportation time to deliver the order.

The next step is to compute the mean and the standard deviation of the demand over the in-bound lead time for each component i, denoted μ_i and σ_i. (The out-bound lead time is used to offset the time-shift in product orders. Hence, if the out-bound lead time for a product is one week, then any planning involving orders for this product will have to be shifted a week earlier.)

We can now write the base-stock level for component i as $R_i = \mu_i + k_i \sigma_i$ just like the model in the last section, with k_i denoting the safety factor. With μ_i and σ_i as parameters derived from given data as outlined above, the decision on the base-stock levels is equivalent to the decision on the safety factors.

Our objective is to minimize the expected inventory budget (capital), subject to meeting the service requirement for each product family. The problem can be presented as follows:

$$\min \sum_{i \in S} c_i \sigma_i H(k_i)$$

$$s.t. \sum_{i \in S_m} r_{mi} \overline{\Phi}(k_i) \leq \overline{\alpha}_m, m \in \mathbf{M}$$

where $r_{mi} := P(X_{mi} > 0)$ is the probability that product m requires component i; c_i is the unit cost of the on-hand inventory of component i; $\sigma_i H(k_i)$ is the expected safety stock of component i, where the H function is defined in (3.4); and $\overline{\alpha}_m = 1 - \alpha_m$ with α_m being the required service level for product family m.

3.4.2 Case Study: IBM's Personal Computer Value Chain

Here we describe a study, which was part of a project aimed at the reengineering of IBM's personal computer value chain from a build-to-stock operation to a configure-to-order operation. To carry out the study, we developed two basic models: an "As-Is" model that is a reflection of the present build-to-stock operation with fixed configurations, and a "To-Be" model that is based on open configurations where the component inventory levels were generated by the algorithm described in the previous section. For both models, we aggregated the production-inventory system into two stages, the first stage consisting of the component replenishment process, and the second stage consisting of the assembly and order fulfillment process. We identified three factors as the focal points of our study:

-   *Manufacturing Strategy.* The "As-Is" operation versus the "To-Be" model.
-   *Forecast Accuracy.* The accuracy of demand forecast at the end product level versus at the component level.
-   *Product Variety.* The effect of mass customization on inventory as a result of direct sales over the Internet.

To study the first factor, we selected a high-volume product family that consisted of 18 finished products that were assembled from 17 components. We used existing bills-of-materials, unit costs, and procurement lead times to develop a detailed simulation model. Demands for each end product were generated statistically based on historical data. The inventory buffers were set to meet a 95% service level requirement for all end products. We then used the product data and the statistically generated demand streams as inputs into the optimization model to determine the optimal base-stock levels for the component inventory. Figure 3.8 shows the comparison between the "As-Is" and the "To-Be" model in the form of overall inventory investment.

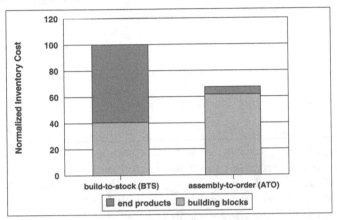

Fig. 3.8. Comparison between build-to-stock (As-Is) and assemble-to-order (To-Be)

To protect proprietary information, the vertical axes were normalized with respect to the inventory investment of the "As-Is" model, which is 100. As expected, the inventory investment for end products was eliminated in the "To-Be" model. (The cost shown is due to WIP; the cost due to finished goods is nil.) The "As-Is" model, in contrast, keeps a significant amount of end-product inventory. On the other hand, the amount of component inventory is higher in the "To-Be" model, which is again expected, since the required service level of 95% is common to both models. Overall, the "To-Be" model reduced the overall inventory investment by roughly 30%. Both models used the same demand forecasts for end products.

In our study of the second factor, we evaluated the effect of forecast accuracy through sensitivity analysis. Figure 3.9 shows the overall inventory investment associated with three different levels of forecast accuracy. The first two columns repeat the comparison shown in the previous figure. The next two columns represent improved forecast errors, at 20% and 10%, achieved by switching to component forecasting in the "To-Be" model.

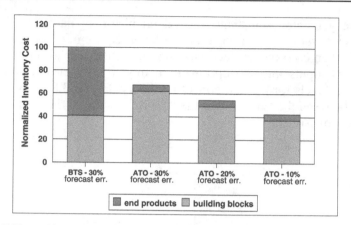

Fig. 3.9. Effect of improving forecast accuracy

Our study of the third factor aimed at analyzing the impact of higher product variety on inventory, with the motivation to support mass customization. In an Internet-based direct sales environment, the number of customer-configured products can be significantly larger than what was supported in the build-to-stock environment with fixed configurations. Figure 3.10 shows the inventory investments.

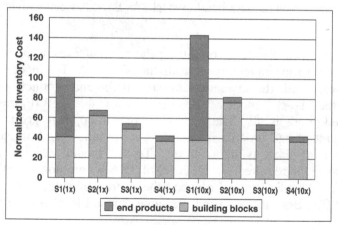

Fig. 3.10. Effect of product variety on inventory

The four columns on the left correspond to the current product set (1x), with scenario S1 as the "As-Is" model, and the other three scenarios as "To-Be" models at the current and improved forecast accuracy levels (scenarios S2–S4). The four columns on the right repeat these scenarios with a product set that is ten times larger in variety (10x). Table 3.1 summarizes all scenarios.

Observe that as the product variety increases, a significantly higher level of inventory is required in the "As-Is" model. This is because forecast accuracy deteriorates when the end products proliferate (i.e., larger varieties result in smaller demand volumes). On the other hand, the inventory increase in the "To-Be" environment is very modest. This is because the proliferation of end products has minimal effect on the forecast accuracy at component level due to parts commonality. This strongly supported the fact that the building-block model is the right process to support a direct-sales operation.

Table 3.1. Summary of scenarios used to study the effect of product variety

	Description	1x Cases	10x Cases
S1	As-Is	Original product set; 30% forecast error; 90% customer service	Ten times larger product set; 30% $\times \sqrt{10}$ forecast error at end product level
S2	To-Be	Forecast at end product level; 30% forecast error; 90% customer service	Ten times larger product set; forecast error as in S2(1x)
S3	To-Be	Forecast at component level; 20% forecast error; 90% customer service	Ten times larger product set; forecast error as in S3(1x)
S4	To-Be	Forecast at component level; 10% forecast error; 90% customer service	Ten times larger product set; forecast error as in S4(1x)

3.5 Semiconductor Value Chains

To meet high levels of service with long lead times and uncertain demands, semiconductor manufacturers often hold large inventories despite the risk and expense. Xilinx, a large manufacturer of application specific integrated circuits, uses design strategies like postponement to better control their inventory expenses (Brown et al. 2000). Under the postponement strategy, inventory is held at in a generic, non-differentiated form in the die bank and is differentiated when the demand is better known. To take full advantage of the postponement strategy, we worked with Xilinx to develop an optimization model that determines the inventory levels in the intermediate and finished goods stocking points to allow for the best service at the lowest cost.

Xilinx, like many semiconductor manufacturers, contracts out most of the manufacturing. Fabrication is performed at vendors in Taiwan and Japan. The wafers are then shipped to assembly vendors in Korea, Taiwan, and the Philippines, where they are held in die bank inventory until needed. Although separate die banks are maintained at each assembly vendor, transfer of die between vendor locations is done as needed. Thus, the die bank is modeled as a single inventory stocking location. Assembled parts are usually shipped to Xilinx facilities in San Jose or Ireland for testing, where they are held in finished goods inventory.

Since die bank inventory is more generic, it serves as an inventory postponement point. The die bank inventory is managed using a wafer starts planning package. Given a desired inventory target, this package determines the total amount to start at the fabrication contractor using a monthly base-stock policy. Finished goods inventories are managed using a planning system based on assembly starts, with separate systems being run in Ireland and San Jose. In the problem formulation below, we discuss how such parts affect the results and how we include such parts in the actual implementation of the model at Xilinx.

As with most inventory decision problems, choosing the appropriate objective function is difficult and highly dependent on the business environment. The key measure of service at Xilinx was total costed "delinquencies" (costed backorders) across all parts, with each part having a different unit delinquency cost. The individual unit delinquency costs were determined using a percentage of unit revenue based on input from the sales department using the following factors: stage in product life-cycle, impact of product availability on future sales of other products, competitive nature of product, and proportion of the product's demand due to key customers. Separate inventory budgets were to be specified for different echelons (die bank versus finished goods) and for different locations within an echelon (San Jose finished goods and Ireland finished goods).

3.5.1 The Optimization Model

The inventory system consists of two echelons. The upstream echelon is die bank inventory and the downstream is finished goods (FG) inventory. The FG inventory is distributed at a set of locations, indexed by $i \in \mathbf{M} := \{1,...,M\}$. Each location i supplies a set of end products, $j \in \mathbf{S}_i := \{1,...,N_i\}$. A separate inventory is kept for each product to satisfy its own demand stream. Let D_{ij} denote the demand (per time unit) for product j at location i, which is assumed to follow a normal distribution. For simplicity, we shall refer to such product at a given location as type ij product.

Let 0 index the die bank location. There are N_0 types of die, each with its own inventory, indexed by $d \in \mathbf{S}_0 := \{1,...,N_0\}$. The relationship between the two stages is a one-to-many mapping: each type of die is used to make one or more end products, but each end product uses only a single type of die. Hence, for each $d \in \mathbf{S}_0$, let $\mathbf{S}_d$ denote the set of end products that use type d die. Then, the demand for type d die is: $\sum_{(i,j) \in \mathbf{S}_d} D_{ij}$. (One unit of end product uses one unit of die.)

Let L_d be the production lead time for each die d. Let L_{ij} be the lead time to transform die d into type ij product. The replenishment lead time for FG inventory (end products) is this nominal lead time, L_{ij}, plus a delay time which takes into account the possible stockout of die bank inventory. This *actual* lead time, denoted $\widetilde{L}_{ij}$, has expected value

$$\mathbf{E}(\widetilde{L}_{ij}) = \mathbf{E}(L_{ij}) + \tau_d p_d, \tag{3.7}$$

where p_d is the stockout probability of type d die inventory used to make type ij product, and τ_d is the expected additional delay when this stockout occurs. Both quantities depend on the inventory levels of type d die. The derivation of the term $\tau_d p_d$ in (3.7) is shown in Brown et al. (2001).

Given information on the demand and lead times, we can derive the means and the standard deviations of demand over lead time: μ_d and σ_d for each die bank location d, and $\widetilde{\mu}_{ij}$ and $\widetilde{\sigma}_{ij}$ for each product ij. Refer to the detailed derivation in De Kok and Graves (2003).

For each ij product, let h_{ij} be the unit inventory holding cost (per time unit), and let s_{ij} be the unit backorder cost. For each type d die, let h_d be the unit inventory holding cost. The decision variables are R_{ij}, the base-stock level for each type ij product's FG inventory, and R_d, the base-stock level for each type d die inventory. As in the last section, we relate these to the safety factors k_d and k_{ij} as follows:

$$R_d = \widetilde{\mu}_d + \widetilde{\sigma}_d k_d , \quad R_{ij} = \widetilde{\mu}_{ij} + \widetilde{\sigma}_{ij} k_{ij} \tag{3.8}$$

and treat the safety factors as decision variables. We have the following optimization problem:

$$\min \quad \sum_{i \in M} \sum_{j \in S_i} s_{ij} \tilde{\sigma}_{ij} G(k_{ij})$$

$$s.t. \quad \sum_{j \in S_i} h_{ij} \tilde{\sigma}_{ij} H(k_{ij}) \le C_i, \quad i \in M;$$

$$\sum_{d \in S_0} h_d \tilde{\sigma}_d H(k_d) \le C_0$$

(3.9)

where C_i denotes the inventory holding cost budget for location i, and C_0 is the inventory holding cost budget for die inventory; s_{ij} is the penalty cost for each unit of backordered ij product; h_{ij} and h_d are unit inventory costs; $\tilde{\sigma}_{ij} H(k_{ij})$ and $\tilde{\sigma}_d H(k_d)$ are expected safety-stock inventories, where H is the function defined in (4); $\tilde{\sigma}_{ij} G(k_{ij})$ is the expected backorders, where G relates to H as follows:

$$G(x) := \mathbf{E}[Z - x]^+ = x - H(x) = \phi(x) - x\overline{\Phi}(x),$$

The objective of the optimization model is to minimize the expected delinquency costs, i.e. the penalty costs of backordered products. To minimize overstocking risk for various products, managers often want to be able to place upper limits on the level of inventory allowed. For example, for a mature product nearer the end of its product life, setting a high inventory target would not be wise. Thus, we add the following constraints on the safety factors to the optimization problem in (3.9):

$$k_{ij} \le \overline{k}_{ij}, \quad j \in S_i, i \in M; \quad k_d \le \overline{k}_d, \quad d \in S_0;$$

where $\overline{k}_{ij}$ and $\overline{k}_d$ are positive upper limits.

3.5.2   Case Study: Xilinx Semiconductor Value Chain

One key question we wished to investigate was how the total inventory holding cost budget $C_0 + C_1$ should be split between the die bank and end products in order to minimize the delinquency cost. Figure 3.11 shows the optimal delinquency cost z^* as a function of $C_0 /(C_0 + C_1)$, i.e., the relative amount of inventory holding cost budget allocated to the die bank, for four scenarios. The inventory holding cost for all die types was held constant at $h_d = 1.0$ across scenarios, but the values of h_{ij} are changed. For the four scenarios we used h_{ij} values of 1.0, 1.75, 5.25, and 10.50 and total inventory holding cost budget of 3,000, 4,000, 8,000, and 12,000 dollars.

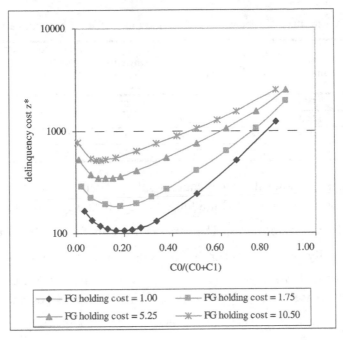

Fig. 3.11. Delinquency cost vs. relative budget allocated to the die bank

From Figure 3.11, we learn that most of the inventory budget should be allocated to end products. For instance, when $h_{ij} = h_d = 1.0$, the optimal operating point is 0.2, suggesting that 20 percent of safety stock should be kept at the die bank, and 80 percent should be kept in finished goods. Table 3.2 shows the delinquency cost z^* and the optimal split C_0^* and C_1^* of the total inventory cost budget for each of the scenarios under study. In additional to reporting the budgets, the table also shows the total safety stock in units allocated to the die bank and finished goods. We observe that the while the percentage of inventory cost budget allocation to die bank actually decreases as the cost of holding FG stock increases, the number of units allocated to the die bank $E(I_0^*)$ versus the number of units allocated to finished goods $E(I_1^*)$ increases. For example, when $h_{ij} = 1.0$, about 20 percent of safety stock inventory should be kept in the die bank, whereas when $h_{ij} = 10.5$ the amount in die bank should be increased to roughly 55 percent. When h_{ij} is large, allocating a larger amount of inventory to the die bank is optimal since multiple units of die can be stocked at the same cost as, say, one unit of finished goods.

Table 3.2. Summary of scenarios used to study the effect of product variety

Holding cost h_{ij}	Total Budget	Opt. DB Budget C_0^*	Opt. FG Budget C_1^*	Objective Value	Die inventory (units) $E(I_0^*)$	FG inventory (units) $E(I_1^*)$
1.00	3,000	600	2,400	106	605	2,423
1.75	4,000	750	3,250	185	755	1,880
5.25	8,000	1,000	7,000	343	1,005	1,349
10.50	12,000	1,250	10,750	519	1,255	1,035

Figure 3.12 shows the percentage of safety stock (in units) allocated to die bank as a function of FG holding cost h_{ij}, for three different values of the FG backorder cost s_{ij}. The amount kept in the die bank increases monotonically with h_{ij}. When s_{ij} is high, the relative amount of safety stock that should be kept at the die bank becomes smaller. For example, in the specific case of $h_{ij} = 10.5$, it is optimal to keep 64 percent die stock when $s_{ij} = 0.875$, 56 percent when $s_{ij} = 1.75$, and 48 percent when $s_{ij} = 4.25$.

In order to project the business improvements that can be achieved with optimization, we compared the optimized policy to the original policy implemented at Xilinx when we began the development of the model. In the original policy, all finished goods buffers (for parts that were build-to-stock) had the same target days-of-inventory, and all die bank buffers had the same target days-of-inventory. We used a full product set in order to compare the original policy and the optimized policy. The product set consisted of 104 die bank parts, 314 finished goods parts held at the Xilinx facility in Ireland, and 1194 parts held at the facility in San Jose. Comparisons between the two policies were made in two different ways. First, the total inventory holding cost budget was held at the same level for both policies, and the projected improvement in delinquency cost was found. Thus, the inventory holding cost that resulted from the original policy was calculated and used as the holding cost budget constraint for the inventory optimization. The ideal split of this inventory holding cost budget between die bank and finished goods was found by running the inventory optimization under a number of different splits. Under the resulting inventory targets, the total delinquency cost was reduced by 54%. The huge improvement was primarily due to reallocation of the inventory among the various finished goods parts. Inventory targets for stable finished goods parts with low delinquency costs and high holding costs were set lower, while inventory targets for less stable parts with higher delinquency costs were set higher.

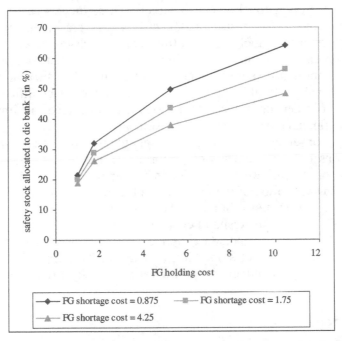

Fig. 3.12. Relative amount of safety stock (in units) allocated to the die bank

Second, the total delinquency cost was held constant for both policies and the projected improvement in inventory holding cost budget was found. Then, the inventory optimization was run with a number of different settings of the inventory holding budget until the budget was found that yielded approximately the same delinquency cost. The optimized policy was found to yield a policy with a 19.9% reduction in overall inventory holding cost (which includes work-in-process) and a 51.2% reduction in inventory holding cost budget when only safety stock was considered.

3.6 Summary

We have described in this chapter optimization models aimed at studying the trade-offs between inventory and service levels in high-technology value chains. We focused on three value chain architectures – complex configured hardware value chains, computer assembly value chains, and semiconductor value chains. Three different inventory optimization methodology based on the characteristics of these three value chain structures were described. Several case studies were presented to illustrate the challenges of the value chains and how the analytical models were utilized to determine the operational inventory targets. These models capture the true multi-tier nature of industrial value chains that are subjected to non-stationary demands, stochastic lead times, and attach-rate uncertainty. The application examples illustrate the importance of not just focusing on the traditional optimal inventory planning and control, but also on exploring alternative value chain designs such as delayed product differentiation, configure-to-order, or supply chain flexibility to achieve maximal operational and financial performance improvements.

We have observed that visibility throughout the supply chain, in particular in terms of measurements of performance and uncertainties, and the flexibility of the business process and infrastructure are critical for the successful implementation of the inventory optimization tools as well as other decision support systems. Some of these issues will be discussed in more detail in the Sense and Respond chapter of this volume. Other issues that warrant further studies include coupling inventory optimization with capacity planning, demand smoothing (via dynamic pricing, for instance), and flexible supply contracts that allow various levels of risk sharing.

References

Brown, A., Lee, H., Petrakian, R. (2000) Xilinx Improves its Semiconductor Supply Chain Using Product and Process Postponement, *Interfaces* 30 (4): 65–80

Brown, A.O., Ettl, M., Lin, G., Petrakian, R., Yao, D.D. (2001) Inventory Allocation at a Semiconductor Company: Modeling and Optimization, in: *Supply Chain Structures: Coordination, Information, and Optimization*, Song, J.-S., Yao, D.D. (eds.), Kluwer Academic Publishers, pp 283–309

Cheng, F., Ettl, M., Lin, G.Y., Yao, D.D. (2002) Inventory-Service Optimization in Configure-to-Order Systems. *Manufacturing & Service Operations Management* 4: 114–132

De Kok, A.G., Graves, S.C. (eds.) (2003) *Supply Chain Management: Design, Coordination and Operation* (Handbooks in Operations Research and Management Science), Elsevier Publishing Company

Ettl, M., Feigin, G.E., Lin, G.Y., Yao, D.D. (2000) A Supply Network Model with Base-Stock Control and Service Requirements, *Operations Research* 48: 216–232

Lin, G., Ettl, M., Buckley, S., Bagchi, S., Yao, D.D., Naccarato, B.L., Allan, R., Kim, K., Koenig, L. (2000) Extended-Enterprise Supply-Chain Management at IBM Personal Systems Group and Other Divisions, *Interfaces* 30 (1): 7–21

Song, J.S., Yao, D.D. (eds.) (2001) *Supply Chain Structures: Coordination, Information and Optimization*, Kluwer Academic Publishers

Tayur, S., Ganeshan, R., Magazine, M., (eds.) (1999) *Quantitative Models for Supply Chain Management*, Kluwer Academic Publishers

Zipkin, P. (2000) *Foundations of Inventory Management*, Irwin/McGraw-Hill, New York, NY

4 Product Pricing in the e-Business Era

Aliza Heching and Ying Tat Leung

4.1 Introduction

This article explores how pricing decisions are made in practice, and how traditional pricing practices are changing in this era of e-business. A survey of the research literature reveals that one of the first explicit calls for model-based pricing decision support appeared in 1978. Today, some 25 years later, we are beginning to see early implementations of commercial price optimization systems of this nature. In this article we provide an overview of common pricing practices and the strategic and tactical pricing-related decisions faced by a seller of products. We survey the existing research literature and discuss the benefits of price optimization. We describe key features offered by commercial pricing systems. Finally, we review some case studies which demonstrate the level of financial benefits that have been derived from the implementation of price optimization systems. The case studies also serve to illustrate the typical first steps taken by businesses that wish to experiment with price optimization.

Product pricing has evolved from simple list pricing, punctuated with an occasional sale or price markdown, to sophisticated pricing mechanisms including auctions, reverse auctions, dynamic pricing, and differentiated pricing based upon factors such as type of consumer and sales channel. The birth of these more sophisticated pricing mechanisms can perhaps be traced back to the time of airline deregulation. Airlines, faced with stiff competition, high costs, and differentiated classes of customers, turned to more advanced pricing mechanisms as a means for financial survival.

The rise in e-business is leading to increased interest by retailers in sophisticated pricing mechanisms. Successful implementation of a pricing mechanism requires a significant amount of data about customers and their buying habits. Traditional (bricks-and-mortar) retailers collect numerous data types daily, including point-of-sale ("POS") purchase data, store traffic data, and logs of customer service calls. Web-based retailers ("e-retailers") have access to another source of data, namely, click-stream data. Click-stream data provides a record of a user's activity on the Web including the order of Web sites or pages the user visits, length of time on each Web page, and possibly email addresses and other personal information associated with the user. Thus, the rise of e-business has brought with it the possibility and profitability of more sophisticated pricing mechanisms in the retail sector. In addition, e-business allows for lower cost and more frequent (if needed) price changes as well as relatively low-cost price testing to gain a better understanding of true market demand.

Product pricing is now a consistent theme of retail trade shows and conferences. Further, an entire industry aimed at providing advanced pricing software solutions has been born, attracting high-tech start-up firms and veterans in supply chain management and enterprise resource planning alike. The growing interest in the successful implementation of pricing mechanisms demands a careful study regarding the effectiveness of such mechanisms. The general press has dedicated detailed articles to this subject; see, e.g., McWilliams (2001), Merrick (2001), Tedechi (2002). A recently published industry study (Marn, Roegner, and Zawada 2003) shows that product pricing is the most effective means for increasing profits among levers including sales volume, fixed costs, and variable costs.

Product pricing mechanisms can be broadly classified into three main categories: products sold through publicly posted prices, products sold through individually negotiated prices, and products sold through auction mechanisms. Two fundamental distinguishing factors between these three mechanisms are which party determines the final selling price and at what point in the sales process is the final selling price determined. In the first category, products sold through publicly posted prices, prices are posted by the seller and are non-negotiable. The purchaser always has full knowledge of the final price that he will pay. Most consumer retail stores in developed countries sell products using publicly posted prices. In the second category, products sold through individually negotiated prices, at the time that the purchaser initiates the buying activity he has no knowledge of the final price that he will pay. As part of the negotiation process, the purchaser and seller engage in extensive discussions regarding many of the contract terms, including the selling price. Subsequent to this negotiation period, the seller determines a final offer price. The purchaser receives the firm price quote from the seller and makes his final purchasing decision. In the final category, products sold through auction mechanisms, at the start of the purchasing activity both the seller and the purchaser have no knowledge of the final sales price. Depending upon the specific auction mechanism, the quantity of product the buyer procures as well as the price per item are revealed only after the purchaser makes a purchase commitment. The final price is determined by the collective set of buyers and sellers participating in the auction. Products sold through responses to Requests for Quotes ("RFQs") put out by a business buyer are often sold using a combination of the second and third pricing mechanisms. The RFQ process is an (reverse) auction; after the winner has been determined, amendments to the originally stated orders (including the selling price) or other forms of negotiation may occur as a result of updated product offerings or changes in the buyer's needs.

In this chapter we focus on the first two categories of pricing mechanisms. We restrict our focus to pricing products that are physical or consumable, such as consumer goods or parts used for manufacturing. We do not consider pricing issues that relate to pricing financial products such as options, or one-of-a-kind artifacts such as antiques or fine art. Finally, we assume that the seller is always a business and do not consider the case of recreational selling of used items or collectibles by an individual.

4.2   Pricing in the e-Business Environment

The traditional bricks-and-mortar business environment is characterized by consumers who must physically enter a store to view merchandise and make purchasing decisions. Retailers face competition primarily from other retailers in close physical proximity. Price change decisions often entail advertising associated with publicizing the new prices. Further, price changes often necessitate a physical marking on each individual item position on the shelf to reflect the new prices. This process is both costly and time consuming. As a result, traditional retailers often limit themselves to a small number of price changes for any given item being sold.

However, the advent of e-business has brought with it some fundamental changes in traditional methods of conducting business. Due to the inherent automation that characterizes e-business, there is a low marginal cost associated with implementing a price change. This low marginal cost allows the seller more flexibility with respect to the number of price changes that he can consider during any given time period. Price changes can often be implemented via a change to a single database entry which will then trigger price label printouts at the retail stores and advertising on the store's website. With the likely widespread future use of either digital display panels or electronic shelf labels, this process will be further simplified.

e-tailing is a "pure" form of retail e-business, where the retailer only has virtual stores on the Web and does not have any physical retail locations. e-tailing is characterized by the use of a website to display products for sale and to receive orders. e-tailing has grown in popularity over the years, as there are many factors that render sales over-the-Web an attractive option for sellers. Displaying products via a website allows e-retailers to build a catalog that is much larger than anything that could fit into a mailbox or into a retail store location. Further, e-tailing allows for significant, if not complete, automation of processes such as order-taking and customer service, thereby reducing transaction costs. Web sales are often characterized by larger purchases per transaction; sellers often dynamically display complementary products to entice additional purchases. e-tailing also provides opportunities for richer interactions with customers, as the use of automated tools allows e-retailers to provide additional services (such as e-mail confirmation when orders are placed or shipped, or when new products of similar kinds are announced) at very low cost to the retailer.

Information technology that enables the existence of e-tailing brings with it changes which impact pricing strategies: (i) The low marginal cost of price changes, as discussed above, allows the seller unprecedented flexibility with respect to the number of price changes and durations of effective prices. Dell.com reports that weekly price changes are routine; in fact, prices can be changed as often as daily (McWilliams 2001). (ii) e-tailing expands the geographic location of customers accessible by retailers. Whereas the reach of retail stores is limited (for the most part) to customers in close geographic proximity to the retail location, websites can be accessed by customers globally. Consequently, product life-cycles (or product shelf-lives) are longer as sellers are not constrained by the seasonal cycles of a single geographic region. The longer selling season impacts pricing decisions used by e-tailers, who now have to consider the larger customer base and more varied customer demands. Further, this expanded reach brings with it a more fragmented market characterized by global competition, as consumers are exposed to websites of sellers from a wide range of geographic locations. e-tailers must now consider pricing actions taken by a potentially large number of competitors and decide whether and how to respond. (iii) e-business increases the number of sales channels via which a seller can reach his customers. The majority of traditional retailers use only in-store sales as a means to generate revenues. (Some retailers also use catalogs as an additional means to access customers.) The growth of e-business introduces new channels by which sellers can access customers, such as shopping from home, kiosks in public places, or even from one's cellular telephone. e-tailers have to consider the role of each of these sales channels, as well as the interactions between them, when making pricing decisions.

4.3   Current Pricing Practice

One can distinguish the use of different pricing mechanisms into two categories, according to the target purchaser of the goods. These two categories are: business-to-consumer ("B2C") and business-to-business ("B2B"). B2C refers to a retailer or manufacturer selling directly to consumers; B2B refers to a retailer or manufacturer selling to other retailers or manufacturers. Table 4.1 provides a list of common pricing mechanisms.

If we consider this list of pricing mechanisms, B2B engagements are most typically paired with special bids (responses to RFQs), auctions, trade promotions, quantity discounts, and annual rebates. Special bids can sometimes be viewed as a special case of a reverse auction where there is only one round of blind bidding. B2C engagements are typically paired with every day low pricing ("EDLP"), high-low or promotional pricing, end-of-season markdowns, bundling discounts, non-linear pricing, non-price promotions, discount coupons, and early bird specials.

Adoption and successful implementation of any pricing mechanism requires both strategic and tactical planning. Strategic planning is used to determine which pricing mechanism(s) to use on what product in which market. Once a pricing mechanism is selected, tactical planning is used to make decisions regarding proper implementation of the pricing mechanism selected during the strategic planning phase.

As an example of this dual-decision process, consider a B2C retailer faced with the strategic decision of whether to adopt an EDLP pricing strategy or a high-low pricing strategy. This decision is dependent upon the target market, the products sold, the long term brand image, and the retailer's overall marketing and operational strategies. Typically, a medium-to-large retailer uses more than one pricing strategy for its different products and markets, and perhaps even for its different channels.

After the strategic decision is made, the retailer is faced with a set of tactical decisions. If the retailer adopts an EDLP pricing strategy, the buyer must determine the single selling price that will be used for the majority of the selling season, as well as markdown decisions for seasonal items (during the end-of-season clearance period) and discontinued items (during the close-out period). If the retailer adopts a high-low pricing strategy, the buyer must determine, for each product, a set of prices that will be used during the selling season. The buyer must coordinate the pricing decisions with non-price promotions decisions. A survey of pricing strategies typically used by retailers of consumer packaged goods can be found in Shankar and Bolton (2003).

Table 4.1. List of common pricing mechanisms

Pricing Mechanism	Description
Special bid	Customized price tailored for each RFQ.
Auction	In its simplest form, public selling of an item to the highest bidder. Many more sophisticated forms exist.
Quantity discount	Price is lowered as a function of the total purchase volume of the order.
Annual rebate	Rebate to purchaser at end of year; magnitude of rebate is determined according to the total purchase value over the entire year.
Contract pricing	Items sold over a given time period at a pre-negotiated price in a pre-specified volume range, possibly with multiple price-volume range pairs. Other conditions such as order or supply lead times also apply.
Trade promotion	Co-operative promotion to the end-consumer by two or more businesses (such as a manufacturer and a retailer).
Every day low pricing ("EDLP")	Item is sold at a single, fixed price; this price does not change over time.
End-of-season markdown	Common practice for seasonal items; reduce selling price at end-of-season in attempt to deplete remaining inventories.
Bundling discount	Price reduction is offered if customer purchases a pre-specified group (bundle) of items.
Non-linear pricing	Different size packs are priced as separate items, not directly proportional to the pack size.
Non-price promotion	Non-price related incentive offered to induce purchase of item (such as positioning of item at prominent locations in a store).
Customer loyalty program	Selected items sold at a reduced price to customers participating in a loyalty program.
Discount coupons	Coupons provided to possibly selected customers that entitle the customer to cash rebates on certain products
Early bird special	Price reduction offered if purchase item during specified time periods.

Estimating demand sensitivity to price and promotions is one of the more challenging aspects of the tactical decision-making process. The buyer often uses the retailer's historical demand and price data to help with this estimation. Typically, the buyer has electronic access to the business' historical data through the use of databases or, more likely, online analytical processing front-ends to databases. For some industries, the buyer may even have historical sales and price data at an aggregate level for a market or product category (e.g., A.C. Nielsen for the grocery industry or A&S for the personal computer industry). Some businesses analyze the impact of promotions and markdowns on sales, typically by estimating "lift factors" corresponding to specific promotion types or markdown percentages used historically in the product family.[4] Market information vendors (e.g., A.C. Nielsen) sell lift factor analysis on commodity products in a set of specific markets (at the aggregate level) or for a specific store (with POS data provided by the customer).

Retailers often use manual or ad-hoc methods, rather than optimization tools, to make these strategic and tactical decisions.[5] For example, many buyers use spreadsheets to compute key performance measures such as total revenue, gross margin, or return on inventory investment for a product family or group of stores over a given time horizon. The buyer uses the spreadsheet to measure the impact of implementing different pricing or promotion decisions on the key performance measures. The results of this analysis guide strategic and tactical decision-making. However, this analysis is time consuming and costly, and the accuracy of the results depends heavily upon the accuracy of any measures estimated by each individual buyer.

In the case of a B2B transaction the same dual decision process is required, but the decisions that must be made are different in nature. For a B2B retailer, strategic decisions include determining criteria for a customer to be eligible for contract pricing, annual rebates or other quantity discounts, and target gross margins for products sold by sales representatives. (These target gross margins may be specified by market or by product family). Other strategic decisions include determining the magnitude of contract or quantity discounts and the value of annual rebates as a fraction of the sales price. Once the strategy is in place, tactical decisions include the degree of control allowed to sales representatives or bid-response

---

[4]   A lift factor measures the change in sales resulting from a price change or promotion, and is computed by comparing the sales volumes between two or more historical time periods which are similar in all aspects except price or promotion type. If more than one aspect of two historical time periods differ, linear regression is typically used to estimate the effect of each factor.

[5]   By the term "manual" we mean that the user makes decisions based upon his estimation. The user may (and most likely will) have access to sources of data, such as historical sales, but these sources simply display historical facts and do not provide predictive computation. We use the term "manual" independent of whether the overall procedure is in any way computerized.

teams.[6] Closely related to these decisions are the incentives offered to the sales teams, which indirectly influence the ultimate selling price. Because these decisions are indirect levers of control, rigorous mathematical modeling is seldom used in practice. Occasionally, a B2B retailer tries to gain insight into optimal tactical decisions by performing empirical studies to compare different regions or to experiment with different degrees of price control.

In a B2B relationship, the B2B seller (e.g., the wholesaler) sells to the B2C seller (e.g., the retailer) who in turn sells to the end consumer. B2B retailers guide their wholesale pricing decisions by estimating demand response to changes in prices and promotions. The wholesaler's attempt to measure end-consumer response to price and promotion decisions is complicated by the following two factors: (i) the retailer employs a pricing and promotion scheme which may not reflect that suggested or employed by the wholesaler and (ii) the retailer may not be willing to share end-consumer data with the wholesaler. To mitigate the effect of these factors, wholesalers often include clauses in contracts with the retailers that define guidelines regarding the relationship between retail and wholesale prices. The impact of (ii) is mitigated by the development and adoption of cost effective information technology and the increasing understanding of the value of information sharing along a supply chain (see, e.g., Gallego (2000)).

For medium-to-large sized B2B businesses, selling price decisions are often left to the sales or bid-response teams. The price for each product sold to each customer is determined based upon a large number of factors including, for example, the previously determined long-term sales strategy for the given customer, the total value of the transaction, the current inventory positions for all of the products in the transaction, and the probability of winning the bid for the transaction. The latter factor is estimated using methods similar to those employed to predict product demand given its price.

The appropriateness of the pricing decisions made by the sales or bid-response teams is largely dependent upon the expertise of each individual pricer. These decisions are generally manually determined, using historical bid or sales data. Prices offered in face-to-face negotiations (as opposed to RFQs) are even more difficult to determine as the pricer must, in general, determine the bid price in real time. Cases where the purchaser provides a yes/no response after seeing the bid price can be viewed as a first-price sealed bid auction (see, e.g., Riley and Samuelson (1981)). In

---

[6]   The degree of control can be expressed as a minimum gross margin, minimum gross profit per transaction, or both.

practice, however, there are often multiple rounds of bidding, even with formal RFQs. This lack of fixed structure in the sales negotiation process complicates the optimal pricing analysis. The pricing decision relies heavily on the potential purchaser's response to the price offered, forcing the practitioner to use a manual process for determining prices.

Thus, optimal pricing and promotion decisions both in the B2C and B2B arenas are difficult to determine. For the most part, these decisions are made using manual techniques, and the quality of the decisions is largely based upon individual pricer expertise and the accuracy of estimates made by the buyers.

4.4 Research Literature

Pricing-related issues have been addressed in the economics, marketing, operations research, and operations management press. In this section we provide an overview of the research literature that can be used for decision support as opposed to papers whose primary contribution lies in studying the dynamics of optimal prices. We refer the reader to Elmaghabry and Keskinocak (2003), Yano and Gilbert (2003) and Chan et al. (2004) for more extensive surveys of existing pricing literature.

Much of the contribution of the economics literature to the pricing area lies in providing high-level models to analyze the various forms of price discrimination, both in B2B and B2C settings. See, for example, Wolfstetter (1999) for a discussion of pricing in a monopoly and an oligopoly. Riley and Zeckhauser (1983) provide an interesting argument describing the benefit to the seller of non-negotiable, posted pricing. Another major thrust of this literature is to understand how changing market conditions impact selling prices. In particular, the literature studies the phenomenon of price "stickiness," where prices remain relatively stable in spite of changes in market conditions. See for example, Blinder (1982) and Amihud and Mendelson (1983). Monroe and Della Bitta (1978) provide a survey on models for pricing decisions, and calls on researchers and practitioners to focus on model-based pricing. The economics literature also focuses on developing models that describe human purchasing behavior. The Bass diffusion model (Bass 1969) is a well-known model for describing how consumers make purchasing decisions. Extensions to this model as well as many additional models of similar nature have been developed in the economics literature.

While the microeconomic models are elegant and insightful, for the most part they do not address operational rules or provide analytical support that can be used in price decision making. We now turn to the contributions of the marketing and operations literature.

B2C pricing has experienced a surge in research activity over the last decade. At the strategic planning level, Ho et al. (1998) study the conditions under which EDLP or high-low pricing is beneficial. In tactical pricing, Smith and Achabal (1998) is one of the first studies that garnered attention from retailers. There, the authors study the problem of pricing during the end-of-season clearance period. They consider a continuous time, continuous price setting with deterministic demand, where the objective is to maximize profit. The retailer must determine the optimal initial (at the beginning of the clearance period) inventory commitment, I_0, and optimal price trajectory as a function of time, $p(t)$. Inventory commitment is defined as the sum of on-hand inventory plus the sum of all future deliveries (orders that have not yet arrived). The sales rate is assumed to be a function of time (seasonality), selling price, and remaining on-hand inventory, and is denoted by $x(p,H,t)$ where p,H,t denote the selling price, on-hand inventory, and time, respectively. The sales rate is assumed to drop when inventory falls below a certain level (referred to as the "fixture fill"). The problem is to maximize profit, given by

$$R(I_0) - c(I_0) = \int_{t_0}^{t_e} p(t)x(p(t),I(t),t)dt + c_e(I_0 - s_e) - c(I_0),$$

where $p(t)$ is the price trajectory at time t, $I(t)$ is the inventory commitment at time t, I_0 is the optimal inventory commitment at the start of the clearance period, s_e is the total number of units sold by the end of the season, c_e is the per unit salvage value for inventory remaining at the end of the season, and $c(I_0)$ is a piecewise linear function that represents total inventory cost. (Total inventory cost includes cost to order additional inventory, cost to display inventory in the store, cost to reduce inventory should the starting inventory commitment exceed the optimal inventory commitment, and salvage value for unsold merchandise remaining at the end of the season.)

The authors use optimal control theory to characterize the optimal solution. By adopting assumptions regarding the form of the sales rate, the authors are able to explicitly solve for the optimal values. The authors conclude the paper by describing three different implementations of their work at major retail chains. The results of these implementations are discussed in the Case Studies section below.

Gallego and van Ryzin (1994) study the stochastic demand version of this model and analyze the problem using optimal control theory. They also extend the problem to cases where only a discrete set of prices is permitted, the initial inventory level is a decision variable, and inventory replenishments are possible (as opposed to a clearance setting where no new inventory will be ordered). Bitran and Mondschein (1997) consider a similar problem, and use dynamic programming to determine the optimal pricing strategy.

Motivated by the work in these two papers, Bitran, Caldentey, and Mondschein (1998) consider the problem of setting optimal prices for perishable products (specifically, fashion items) at a chain consisting of n stores. They work together with a retail fashion chain in Chile to develop markdown pricing policies, and compare the results of these policies to those achieved by the retailer. The items have little value at the end of the season, as they are fashion items that are not in demand once the season has ended. The problem is to optimally set prices during the clearance markdown period, with the objective of maximizing total discounted expected revenue. The retailer is constrained to charge a common price at all retail locations, though the demand distributions are not necessarily the same across retail locations.

The selling season is divided into K review periods, where pricing decisions must be made at each review period. (Due to the short selling season relative to the long inventory procurement lead time, it is assumed that inventory decisions are made prior to the start of the selling season and no inventory replenishments are possible during the selling season.) The customer arrival process at each retail location follows a Poisson distribution with a time-dependent arrival rate, $_{ik}$, where $i=1,...,n$ denotes the store location and $k=1,...,K$ denotes the review period. (We note that review periods are numbered according to time-to-go, so that $k=1$ denotes the end of the season.) Customers will only purchase an item (i.e., an arrival will be converted to actual demand) if the item's selling price, p, is below the customer's reservation price. The authors assume that demands for different items are uncorrelated (though they mention that revenue improvements could be realized if demand correlation is explicitly considered in the model). The paper considers two different business rules: (i) no inventory transshipments between retail locations are permitted and (ii) inventory transshipments between retail locations are permitted during the selling season. The authors formulate the problem under each of these business rules using dynamic programming. We present here the model assuming business rule (i), where no inventory transfers are permitted, is followed. Thus, the problem under business rule (i) is formulated as follows:

$$V(c_{1k},...,c_{nk}) = \max_{p \geq 0} \sum_{j_1=0}^{\infty} \cdots \sum_{j_n=0}^{\infty} [p(\min(c_{1k} \cdot j_1) + \cdots + \min(c_{nk} \cdot j_n) +$$

$$V_{k-1}(c_{1k} - \min(c_{1k}, j_1),...,c_{nk} - \min(c_{n1k}, j_n))] \cdot \Pr(D_k(p) = j_1,...,j_n),$$

and boundary conditions

$$V_k(0,...,0) = 0 \text{ for all } k=1,...,K \text{ and}$$

$$V_0(c_{10},\ldots,c_{n0})=0 \text{ for all } c_{10},\ldots,c_{n0},$$

where $V_k(c_{1k},\ldots,c_{nk})$= total expected revenue from review period k until the end of the selling season, when starting inventory at retail location i is c_{ik} and the optimal price is implemented in each period, for all $i=1,\ldots,n$ and $k=1,\ldots,K$, and $Pr(D_k(p)=(j_1,\ldots j_n))$ = probability that $j_1\ldots j_n$ customers (in store locations $1,\ldots n$, respectively) wish to purchase the product at review period k, for all $k=1,\ldots,K$.

Due to the large dimension of the state space, for any reasonable value of n, the authors propose heuristic methods for solving the problem. The authors conclude by applying their methodology to real data obtained from a fashion retail chain in Chile and comparing the profits achieved to those actually achieved by the company. The authors studied six products, but report on the results of two. (The remaining four products had similar results.) For product 1, the heuristic leads to a 16.1% increase in revenue over those actually achieved by the chain; the results for product 2 are similar.

Tellis and Zufryden (1995) consider a more comprehensive demand model which includes the effects of brand loyalty, stockpiling, and customer segmentation. The profit maximization problem is formulated as a nonlinear integer program and is solved using the Solver optimization module in an Excel spreadsheet. While this approach can provide insight into the more general pricing problem, it is not a practical solution for a retailer with tens or hundreds of stores and thousands or tens of thousands of items in each store.

Heching et al. (2001) consider the case of maximizing revenue or profit over the entire selling season (of length T periods) for custom configured products with related bill-of-material structures. The custom configured products are composed of a number of components. The retailer faces periodic decision points; at each decision point the retailer must determine, for each component, the procurement quantity and any promotions or markdowns to offer in the current period. Demand, however, occurs at the system level (consisting of a set of components as specified in the bills-of-material). The selling price of a system is simply the sum of the prices of the components.

Demand for component b at time t is modeled as a deterministic function defined as

$$D_{bt}(m_{bt},p^r_{bt}) = f(s_{bt},m_{bt},p^r_{bt},m_{b't},p^r_{b't}),$$

where

m_{bt} = markdown for building block b at time t, t=1,…,T

p^r_{bt} = promotion offered for building block b at time t, t=1,…,T

s_{bt} = seasonality factor for building block b at time t, $t=1,\ldots,T$

$m_{b't}$ = markdowns for related building blocks b' at time t, $t=1,...,T$
$p^r_{b't}$ = promotions for related building blocks b' at time t, $t=1,...,T$.

The final two parameters capture the impact of other components on demand for component b. For example, if building block b represents a 16MB memory chip, promotions or markdowns offered on 32MB memory chips may impact the demand for the 16MB chip. Promotions and markdowns are selected from a pre-specified menu of options.

The demand models for the different components are constrained to ensure that the ratios of these component demands satisfy the final products' bills-of-material. The retailer faces both promotion and markdown budgets; the total dollar value of all sales-related promotions (markdowns) for all components combined cannot exceed the promotion (markdown) budget. The promotion budget constraint can be expressed as $\sum_{t=1,...,T} \sum_{b \in B} \sum_{p^r \in P^R} c_{bp^r t} \cdot u_{bp^r t} \leq C^{P^R}$, where $c_{bp^r t}$ is the cost for offering promotion type p^r for building block b at time t, $c_{bp^r t} = 1$ if promotion p^r is offered for building block b at time t and $c_{bp^r t} = 0$ otherwise. The following variables are defined to specify the relationship between the pre-specified menu of allowable markdowns ($m \in M$) and the actual percentage reductions in selling price from the list price for component b announced at the beginning of the selling season (p_b). Let $v_{bmt} = 1$ if markdown m is offered for building block b at time t and $v_{bmt} = 0$ otherwise. Assuming that markdowns are offered in 5% increments, the magnitude of the actual markdown offered for component b at time t is defined by $m_{bt} = \sum_{m \in M} 0.05 \cdot (m-1) \cdot v_{bmt}$ where $m=1$ corresponds with no markdown. The constraint $\sum_{m \in M} v_{bmt} = 1$ ensures that only a single markdown is offered for any component in any time period. Then the markdown budget constraint can be written as $\sum_{t=1,...,T} \sum_{b \in B} \sum_{m \in M} p_b m_{bt} v_{bmt} \leq C^M$. Similarly, the retailer faces a procurement budget such that the total amount spent procuring components cannot exceed the procurement budget. One can incorporate various business-related constraints such as a limit on the total number of markdowns or promotions offered (to maintain their effectiveness), restrictions relating to markdowns or promotions in consecutive periods, and constraints that restrict the retailer from offering a markdown or promotion during the first few periods of the selling season.

The objective is to maximize total profit over the entire selling season; the objective function can be written as follows:

$$\max \sum_{t\in1,...,T}\sum_{b\in B}\{revenue_{bt} - procurement\ \cos t_{bt} - promotion\ \cos t_{bt} - lost\ sales\ \cos t_{bt}\},$$

where

$$revenue_{bt} = q_{bt} \cdot p_b \cdot (1 - m_{bt}) = p_b \cdot q_{bt} - 0.05 p_b \sum_{m\in M} (m-1)v_{bmt},$$

$$procurement\ \cos t_{bt} = c_{bt} \cdot q_{bt},$$

$$promotional\ \cos t_{bt} = c_{bp^rt} \cdot u_{bp^rt},$$

$$lost\ sales\ \cos t_{bt} = \pi_{bt}(d_{bt} - q_{bt}),$$

q_{bt} is sales of building block b at time t, c_{bt} is the cost to procure a unit of component b at time t, π_{bt} is the per unit stockout penalty cost for component b at time t, and d_{bt} is demand for building block b at time t.

(We note that in the case of supply constraints, demand in a given period may exceed sales in that period.)

The difficulty in the formulation arises from the fact that the revenue term is non-linear in both $q_b t$ and v_{bmt}. Considering that the number of feasible solutions is exponential in the length of the planning horizon (T) and the number of allowable markdowns, the possibly high frequency with which some retailers review and dynamically modify product prices (see, e.g., McWilliams (2001)), and the large number of components interrelated through the bill-of-materials, this optimization problem cannot be solved in any reasonable amount of time for a typical length of planning horizon and set of permissible markdown. The authors develop efficient heuristics that provide close-to-optimal solutions while ensuring a reasonable run-time.

Gallego and van Ryzin (1997) consider a similar problem of maximizing the revenue from a set of products over a finite horizon, assuming that the product demand follows a stochastic point process. The authors first analyze the deterministic version of the problem and show that the deterministic model provides an upper bound on the revenues that can be achieved. Based on the solution to the deterministic model, the authors suggest heuristics for solving the stochastic problem. These heuristics are shown to be asymptotically optimal as expected sales tend to infinity. Heching et al. (2002) report on an empirical study in which results from such optimization models are compared to the pricing decisions made by a retailer. Their results indicate that retailers should take markdowns of smaller magnitude earlier in the season, rather than taking steep markdowns late in the selling season. (See the Case Studies section for more detailed discussion of this work.)

Sometimes, for each product, there exists a menu of fixed prices from which the planner can select. Such situations can arise when pricing and product planning functions are performed by different organizations within a company. Feng and Gallego (1995) consider this problem under a Poisson demand assumption, where the planner must decide the timing of a single price change (either price increase or price decrease). The authors use optimal control theory to show that there exists a sequence of price-dependent time thresholds that dictate whether or not a price change should occur given the quantity of inventory remaining at any point in the sales horizon. Aviv and Pazgal (2003) also consider the case of a single price change. Here, the authors consider a case where demand is correlated over multiple periods. The retailer must make three decisions: an initial selling price (p_1), a markdown price (p_2), and the time at which to make the price change (T_D). Customers know the nature of the pricing policy (i.e., single markdown) but do not know the values of (p_1, p_2, T_D). The authors study the loss in profit that arises when sellers assume that consumers are myopic (i.e., purchase at time t if the selling price is below their reservation price, ignoring the knowledge that a price markdown will occur in the future).

Aviv and Pazgal (2002) consider dynamic pricing of fashion goods over a short selling season. The retailer is uncertain of the exact value of the parameters of the demand distribution, and it modifies its estimate of the demand distribution as he observes demand.

Closely related to the problem of price optimization is the combined problem of determining price and inventory levels. Eliashberg and Steinberg (1991) provide a survey of problems that lie at the interface between marketing and production decisions. More recent works in this area include Subrahmanyan and Shoemaker (1996), Federgruen and Heching (1999), Petruzzi and Dada (1999), and Van Mieghem and Dada (1999). Chan, Simchi-Levi, and Swann (2002) consider combined pricing and inventory control in the presence of capacity constraints, lost sales, and discretionary sales, where a retailer may choose not to meet all demand even if inventory is available. Chen and Simchi-Levi (2002) allow for a fixed component to the ordering costs. A recent paper by Ahn et al. (2004) considers combined pricing and inventory decisions when the purchase decision in period t is dependent on prices over multiple periods rather than the price prevailing during period t alone. Also related is the problem of pricing products in conjunction with service-related decisions. See Hassin and Haviv (2003) for a survey of basic models in this literature. Extensions to more complicated situations have been suggested, for example, by Bernstein and Federgruen (2001) and Maglaris and Zeevi (2003).

Many of the papers referenced above consider a setting where sellers operate as monopolists. There has also been significant research interest focusing on pricing decisions in the face of horizontal or vertical competition. This area of the literature assumes that sellers may be facing external competition and may also be managing a portfolio of competing products.

See, e.g., Gallego and van Ryzin (1997), Tsay et al. (1999), Gilbert (2000), and Zhu (2002).

The B2B pricing research literature is less rich than its B2C counterpart. Papaioannou and Cassaigne (2000) provides a recent review of statistical models for bid pricing in an RFQ environment. A basic assumption in these earlier models is that complete historical data on bids (including those submitted by competitors) are available. This assumption is satisfied for the purchaser, but not for the seller. To avoid this problem, Cassaigne and Papaioannou (2000) proposed an expert system approach to estimate the bid-win probability (i.e., the probability that a seller will win a bid). Similar in spirit, but using a data mining approach, Lawrence (2003) estimates the bid-win probability using only those data available to the seller. Cao et al. (2002) use a machine learning approach to determine the win probabilities and to estimate missing win-loss information from historical bidding data. One could also use discrete-choice analysis to model buyer behavior and to estimate the bid-win probability. See, e.g., Ben-Akiva and Lerman (1985) for a discussion of discrete choice models. Talluri and van Ryzin (2000) have used this approach in the context of airline revenue management.) Once the bid-win probability is estimated as a function of selling price (and other factors), the problem of maximizing the expected profit of that particular bid is relatively straightforward.

Another stream of B2B literature focuses on the relationship between price and promotion decisions taken by the wholesaler and the retailer. In a number of industries, manufacturers often plan promotions (both price and non-price related) in collaboration with retailers. In these cases the manufacturer typically contributes some money to an end- consumer promotion, say, in the form of a direct payment or a price reduction to the retailer. The retailer may then decide to contribute his own money to boost the promotion, for example, in the form of a price reduction to the consumer. Alternatively, the retailer may decide to retain the entire promotion contribution from the manufacturer and take no action to promote the product to the consumer (though such actions are sometimes restricted by terms and conditions in an explicit contract between the wholesaler and the retailer). The fraction of a manufacturer's promotion that is reflected in a promotion seen by consumers, called the pass through rate, is a retailer's decision that can be optimized. Arjunji and Bass (1996) describe a model to optimize the pass through rate, retail promotion duration, and order quantity for a manufacturer-promoted product. Krishna and Kopalle (2003) investigate a similar situation in a multi-product environment. Silva-Risso et al. (1999) report a decision support system for a manufacturer to determine an optimal promotion plan given a known and constant pass through rate. At a more strategic level, Neslin et al. (1995) investigate the relationship between retailer / consumer behavior and the manufacturer's optimal promotion plan. Although the focus of the paper is on managerial insights, the optimization model described therein provides guidelines for a model that could be used for tactical decision support.

4.5 Commercial Systems

Though airlines have been profitably employing sophisticated pricing mechanisms (yield management) for over two decades, retailers have been slower in adopting these more advanced methods. Retail pricing decisions have traditionally been left in the hands of buyers, who rely on a combination of intuition and spreadsheet calculations to make pricing decisions. Decisions are often driven by target margin objectives, frequently resulting in misalignment between consumer demand and retail prices. However, successful implementation of yield management in airline pricing as well as tougher economic conditions have convinced retailers that there may be financial benefit in using mathematical models for optimizing pricing decisions. This growing recognition has brought with it a demand for solution providers to develop software that addresses the complexities associated with retail pricing optimization.

In response to this demand, a number of software tools have been developed with the objective of improving retailer profitability through price optimization. In this section we discuss the available commercial price optimization tools. At this time, the majority of commercial systems are designed to support decision-making in the B2C retail industry with publicly posted prices. For brevity, we use the term price optimization system with the understanding that the system may also provide promotion optimization.

Most of the vendors who offer retail pricing optimization tools are new to the revenue management arena, and have not traditionally offered airline yield management tools. These include DemandTec, i2 Technologies, Khimetrics, KSS Group, Manugistics, Metreo, ProfitLogic, Rapt, Spotlight Solutions, Zilliant. O'Neill, Daggupaty, and Cauley (2003) and Elmaghraby and Keskinocak (2002) provide an overview of some of these vendors. Supply chain management vendors, such as i2 Technologies and Manugistics also provide offerings in the price optimization area.

The commercial offerings typically provide two major functions: (i) a demand model and (ii) price optimization. The demand model determines demand as a function of selling price and other factors. This demand function is used by the optimization model to maximize profit or revenue, while considering user-defined constraints such as business rules, current inventory levels, required service levels, and length of the selling season. The business rules constraints ensure that the computed solution is sensible from the end-consumer's perspective and that specified business strategies and policies are observed. For example, the seller may constrain the system such that a larger package size of a product should be priced higher than a smaller package size of the same product, or that national brands should be priced at least as high as a house brand of the equivalent product. Other business rule constraints may include the number, magnitude, or frequency of allowable markdowns, or constraints requiring that groups of items must always be marked down simultaneously. An additional feature offered by these systems is to consider the multiple sales channels (and multiple store locations within the "bricks-and-mortar" sales channel) and provide optimal channel-specific and location-specific prices for each product.

In developing demand models, each software solution from a vendor uses its own, often proprietary, method. This difference can be quite critical, as the estimated demand function is an underlying driver of the price optimization model. The form and coefficients of the demand model are determined using historical sales and price data. Cost data, competitive actions, prevailing market conditions, cost of capital, salvage values, and inventory carrying costs are also important factors to be considered. Ideally, historical sales data are obtained from corporate databases or directly from POS systems. Methods for modeling demand include, for example, simple "lift factor" calculations, traditional econometric models, and consumer choice models. Some vendors determine the appropriate demand model for each product by using an "attribute management system." In this approach, products with similar attributes are clustered together. A library of demand functions is maintained, and econometric modeling is used to find the demand function that fits best with each cluster of products.

Developing demand models and searching for a revenue or profit maximizing solution given these demand models (with estimated parameters) and the business constraints, are nontrivial tasks with respect to computational complexity and the quality of the solution. These two tasks serve as technical differentiating factors in the business. To specify the constraints, most vendors provide a user-friendly interface. For example, a list of related constraints can be specified by using a 'for' loop, similar to a high level programming language. Managing these constraints is challenging since there is typically a large set of constraints (often in the thousands) which need to be manually input and maintained. Even if one considers a simplified demand model where each product is modeled independently of other products, many business constraints (such as the relationship of the prices of the different pack sizes) will link products together, producing a large set of constraints.

In the retail trade, the prices generated by the price optimization system are reviewed by the buyer. Buyers often perform scenario testing to study the profitability of implementing the suggested pricing strategy under different scenarios. (Most commercial systems offer some level of automated scenario testing capability.) Once the buyer determines the final pricing strategy, the prices must be rolled out to store locations. Some automation method has to be put in place, such as a bridge between the price optimization and price management or POS systems. The results of these price decisions must be measured and monitored as consumer response to retail prices is observed, to ensure that the planned profit or revenue will be met. To this end, the price optimization software may have functionality that allows the retailer to analyze and monitor the impact of pricing decisions on sales and margins. Price adjustments due to competitive actions and seasonal changes may require the retailer to use the price optimization system one or more times to revise the retail prices.

Recently, a small number of systems have been developed to support pricing in B2B arrangements. For example, the B2B pricing system developed by Manugistics analyzes a specific customer contract (for example, a contract proposed in the context of an RFQ) and recommends optimal prices for the set of products included in the contract. The logic is fundamentally similar to that of a B2C system with the exception that each customer is classified into a specific market segment and historical data from that segment alone is used to estimate the demand model. In addition, a contract-win probability is estimated as a function of price and other factors.

The demand model generated by the price optimization system is used to drive pricing decisions. On the other hand, businesses typically have a demand forecasting system, perhaps appearing as a module in their enterprise resource planning or supply chain management systems, that is used to drive inventory, production, and other planning decisions. The use of different demand forecasting models to drive different business decisions is highly undesirable and may lead to uncoordinated decision-making. Consequently, methods must be determined to reconcile between the forecasts generated by the two systems.

It should be noted that the terms "price or promotion optimization" or "price or promotion planning" have been used rather loosely in the marketplace. In some cases the system does not provide any automatic optimization per se, but instead provides relevant information (such as historical sales reports) that help the user optimize prices or promotions. These systems do not have an underlying demand model or an optimization engine, and are instead focused on business data analysis, data management, and workflow. OLAP (On-Line Analytical Processing) based systems or price management systems (e.g., price management modules in common ERP systems) fall within this category. Such systems are clearly useful in their own right but are not the focus of this article.

4.6 Benefits of Price Optimization

As with revenue management systems used by airlines, it is difficult to accurately assess the financial benefits of a price optimization system. The accuracy of this estimate depends largely upon the accuracy of the estimate of demand sensitivity to prices and promotions, which is difficult to measure. Typically, one of the following two approaches to estimate a demand model is adopted:

(i) The seller uses historical data to develop a demand model. The seller uses this demand model to simulate historical sales (and associated profits and revenues) assuming that the prices (and promotions) suggested by the price optimization system are adopted. The profits and revenues generated in the simulation are compared with the true historical profits and revenues. This gives an estimate of the profit and revenue improvement that can be achieved by using the price optimization system. The estimated improvement is adjusted to account for inaccuracy in the demand model. The adjustment is commonly performed by estimating the inaccuracy in the demand model in one of two ways: (a) Compute the percent difference between the demand predicted by the demand model using the historical price vector to the actual historical demand. This percent difference is used to adjust the revenue improvement estimated in the simulation. (b) Compute the percent difference between the demand predicted by the demand model using the historical price vector and the demand predicted by the demand model using the price vector suggested by the price optimization system. This percent difference is used to adjust the revenue improvement estimated in the simulation. The intuition behind this method is that the predicted differences in demand (when the prices are different) may be relatively accurate, even though the actual demand observed for any given price may not be.

(ii) A potentially more costly but perhaps more convincing method for measuring the benefit of price and promotion optimization is to conduct a pilot study. For example, in the case of a retail store chain, a subset of retail locations adopts the price and promotion strategy suggested by the price optimization tool. Profits generated by this subset of retail locations are compared with the profits generated by the control set of retail locations for which traditional pricing rules were applied, to measure the benefit of the price optimization system. This approach eliminates the direct dependence of the estimate of the benefit on the accuracy of the demand model. On the other hand, for accurate measurement of benefit, the retailer must find two comparable, representative, and sufficiently large sets of retail locations, and ensuring that there are no unusual factors or events that occur during the time of the experiment. Perhaps the greatest hurdle is that the retailer must have sufficient confidence in the price optimization system to conduct such an experiment in a significant number of stores over a reasonably long period of time.

In practice, both approaches are used for estimating the benefit of price optimization systems. As the price optimization industry matures and an increasing number of successful implementations of price optimization systems is known, more businesses will gain enough confidence to conduct pilot studies.

Because the price optimization industry is very young, the long-term value of such systems is yet to be established. Further, the magnitude of the financial benefit depends on the particular business environment and the method of implementing the price optimization system. However, a number of pilot studies in the retail industry have been published and their results are encouraging. Improvements in revenue on the order of 1-5% in the pilot implementations have been reported. The associated improvement to the bottom line is often significantly larger. Feldman (1990) reports that for an industry with a 1.6% profit margin, a 1% revenue improvement translates to a 60% increase in profits. Other quantifiable benefits include reduction in inventory levels (especially for seasonal products) or, equivalently, an increase in sell through, improvement in gross margin return on inventory investment, and reduction in labor costs due to a reduction in the number of unnecessary markdowns. See, e.g., Johnson, Allen, and Dash (2001), Girard (2002), and Scott (2003), for discussions of actual implementations of price optimization systems and the benefits observed in those cases. Our own experience (further described in the Case Studies section) is consistent with these published results.

The literature relating to B2B pricing contains a void with respect to information on pilot studies or practical experience in implementation of price optimization systems. (We ourselves do not have any direct experience in developing such systems for a B2B enterprise.) This phenomenon, coupled with the observation that the B2B pricing research literature is much less active than its B2C counterpart, indicates that significant opportunity exists for additional work and development in this area.

A common question is how a mathematical model, relying primarily on historical sales data, can outperform an experienced retail buyer or product pricer and can produce such significant financial gains. There are two arguments to support this phenomenon.

First, a medium to large business (at present the target user of such systems) has many buyers or pricers with varying degrees of expertise. While the price optimization system may not outperform the more experienced pricers, it will be helpful to the less experienced ones. The business, as a whole, therefore benefits. This point should be noted when a business is selecting pricers for a pilot study of a price optimization system. Some businesses may be inclined to include only the top buyers or pricers in a pilot study and conclude that the price optimization system is not beneficial because it does not outperform the top people. The business should consider the broad range in expertise of its pricers when assessing the benefit of a price optimization system.

Second, even more experienced buyers or pricers have difficulty performing well at the store-product (in B2C) or customer-product (in B2B) level. A medium to large sized business has a large number of store-product or customer-product combinations, each of which might have very sparse historical data useful for pricing decisions. Further, it requires significant time for the pricer to analyze every store-product or customer-product combination to make good pricing decisions. As a result, buyers or pricers are often forced to adopt potentially suboptimal decisions such as a common price for a product over all store locations, or a common discount rate for all products for a customer or customer-segment. On the other hand, processing a large number of items with detailed data is precisely the strength of computer-based models. The price optimization system can price each store-product or customer-product differently, based upon the historical behavior observed at each store or customer. Thus, while an experienced buyer or pricer may be able to more accurately predict aggregate behavior of a product family for the entire business, the price optimization system can often more accurately predict behavior at a detailed level.

Both of these arguments support the concept that a decision support system such as a price optimization system complements the ability of its human user. For example, the buyer may be more accurate at determining the demand trend for a product family at the retail chain level, and the price optimization system can be used to compute the demand models at the store-product level, given the high-level demand trend specified by the buyer. Also, the buyer may be more accurate in predicting the demand trend for large product families or products that appeal to specific customer types. The buyer can be used to determine the pricing scheme for these products and the price optimization system to estimate the demand and determine prices for the other products. In these ways, the buyer's time can be more efficiently utilized. It allows the buyer more time to analyze other, more qualitative though equally important factors (such as fashion trends) or competitive behavior. By combining expert knowledge with a data-based optimization model, it is more likely that a business can see significant improvements in pricing performance and in overall profits.

4.7 Case Studies

In recent years, many business, especially those in the retail industry, are attempting to take advantage of price optimization systems. The following retail businesses have been reported to have tried or be in a pilot program or be in some stage of implementation: Casual Male (Agosta 2001; Boone 2002), D'Agostino Supermarkets (O'Neill et al. 2003), Gymboree (Agosta 2001; Merrick 2001), JCPenney (Agosta 2001; Girard 2002; O'Neill et al. 2003; Merrick 2001), KB Toys (O'Neill et al. 2003), Longs Drug Stores (Girard 2002; O'Neill et al. 2003), Saks (O'Neill et al. 2003), and ShopKo (Girard 2002; Johnson 2001; O'Neill et al. 2003; Merrick 2001).

In particular, ShopKo reported an estimated sales increase of 14% and corresponding gross margin increase of 24% for 300 items involved in a markdown optimization pilot study (Johnson 2001; Merrick 2001). Casual Male, a specialty apparel retailer for big and tall men, is in the process of rolling out a chain-wide implementation of a markdown optimization system (Boone 2002). This decision followed a successful pilot conducted in the fall of 2001. The pilot lasted about three months, during which a markdown optimization system was used in six of the 25 departments in all of its anchor stores. (This represents approximately 25% of the Casual Male's business.) A few years of sales history was used to initially populate the database and estimate the products' price elasticities. During the pilot, the database was refreshed weekly with new sales data. Buyers at Casual Male accessed the markdown optimization system remotely to review the system-recommended prices to maximize gross profit margin while selling the inventory by a target date, and to perform what-if analyses. Although specific performance metrics from the pilot were not disclosed, the improvements in gross margin and sell-through achieved in the departments that used the markdown optimization system met or exceeded Casual Male's expectations. It is also interesting to note that the software vendor hosted the markdown optimization system for both pilot as well as production use.

In the computer industry, Dell is reported to have at least experimented with price optimization systems, and Hewlett Packard in 2002 attributed a $15M increase in revenue due to the use of a price optimization system (Tohamy 2003).

The research literature has also reported evidence of successful applications of such systems in retail settings. Most notably, Smith and Achabal (1998) discuss implementations of their proposed clearance pricing methodology at three major retailing chains. The implementations required the adoption of various assumptions and required some subjective estimates of some of the model parameters. In spite of these estimates and assumptions, two of the implementations were successful, resulting in increased revenues for the chains. In the third implementation, the model's performance was inferior to the chains' existing pricing policies. The successful implementations included a mass merchant retailer with over 600 store locations and a mid-market department store chain with approximately 300 store locations. The estimated benefits were annual profit increases of over $10 million and $15 million, respectively. In both cases, the computed optimal prices served as decision support for the buyers; in one case the computed optimal prices were approximate because only a subjective estimate of the price sensitivity parameter was available. The unsuccessful implementation occurred at a large general merchandise chain with approximately 800 store locations. Test results there indicated an increase of 5% in clearance markdown dollars using the model when compared to existing practice. The authors attribute the unfavorable results to lack of good inventory data, poor estimate of demand sensitivity to promotions, and the fact that prices changed weekly while sales data were reported monthly, leading to an inability to correctly adjust the parameters in the model.

Leung and Ramaswamy (1998) conducted a markdown feasibility study for a department store chain. At the time that this feasibility study was conducted no commercial price optimization software existed, though a few boutique firms were offering technical price optimization consulting. To conduct a real pilot program (as described in approach (ii) in the Benefits of Price Optimization section), a retailer is required to custom-develop a costly price optimization system. This feasibility study was conducted to serve as a low-cost means to measure the potential benefit of implementing a full price optimization pilot. Subsequent to the completion of the feasibility study, the retailer embarked on a pilot program to test the effectiveness of implementing price optimization.

For the purposes of this feasibility study, the authors implemented a published price optimization algorithm but developed a simulation model around the optimization algorithm. (See details below.) The objective of the study was to assess the potential revenue improvement by adopting regional or store-unique price markdown planning using a price optimization system.

The sample data set used for this study contained 38 weeks of sales (and price) data for a family of women's apparel items from 85 store locations, and accounted for approximately $20 million in actual revenue over the time span considered. The items included in the family belonged to the same product type and had historically been marked down identically at the same time (e.g., 10% off for every item in the family). To retain this practice, the authors characterized all the items by a single demand model, which effectively represented the aggregate behavior of these items. This single demand model was used to drive the optimization algorithm, giving a single recommended markdown. Each store location, however, might be modeled separately to enable different markdowns at different locations. The steps of the feasibility study were as follows:

1. Sales data was aggregated by location so that each location has a single set of price-demand data (representing an "average" item in the family). The highest price during the 38 week period was assumed to be the regular or list price for the aggregate. Using this price, markdowns were calculated for each week, and for each location.
2. The set of sales and markdown data was then clustered (Leung et al. 2002) to isolate groups of locations that exhibited similar price sensitivity and seasonality.
3. From each of the three resulting clusters, a small number (3-4) of locations were selected at random and a model of demand as a function of price, base sales rate, and seasonality was fitted to the grouped data. A total of eleven store locations were included.
4. Using this demand model and the price markdown algorithm published in Smith and Achabal (1998), a simulation over the 38 week period was carried out. Demand variability was approximated using a Poisson distribution (with the mean given by the fitted demand model). One hundred replications were averaged to estimate the average revenue from the algorithm. Due to restrictions imposed by the retailer, no price changes were allowed for the first ten weeks of sales. The revenue over the following 28 weeks was compared to the actual, realized revenue for that location and the percentage improvement calculated.

Key observations from this exercise are as follows:

1. The 85 stores could be classified into one of three groups – high, moderate, and low price sensitivity, corresponding to a demand lift of 14%-25%, 10%-14%, and less than 10%, respectively, for a 10% price markdown.

Actual price markdown practices were on-target only for store locations with a price sensitivity that was in the middle of the high price sensitivity range. For the remaining store locations, significant increases in revenue could be realized by better managing the timing and magnitude of price changes. For the sample of selected store locations in each cluster, simulation experiments were run to reveal revenue improvement potential as shown in Figure 4.1. The revenue obtained by using the price optimization algorithm was computed assuming that the demand realized would follow that predicted by the demand model. In practice, the demand model is not completely accurate and the actual revenue increase is less than that predicted by the simulation. As a conservative estimate, the authors recommend reducing the estimated revenue improvement by 50%. In this case, the result was an estimated 4% increase in revenue.

2. The stores' estimated price sensitivities were (partially) validated by a panel of store executives who examined each store location relative to the demographics of the customers in the area. For example, stores with low price sensitivities were consistently located in posh shopping districts of metropolitan areas.

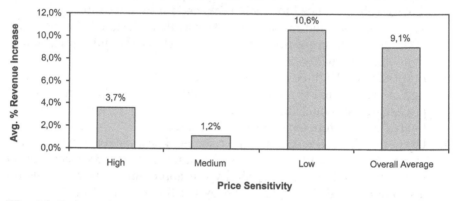

Fig. 4.1. Estimated Revenue Improvement in Customer Case

The results indicated excellent potential for substantial improvements in profitability by the ongoing use of a price optimization system. The operational costs of a possibly greater number of price changes than is the current practice, legal and other constraints that limit price variations between locations, and maintaining the price optimization system itself all need to be weighted against the benefits. But the scale of the estimated improvement appeared sufficiently large to more than accommodate these factors.

Heching et al. (2002) describe another dynamic pricing feasibility study that was performed for a fashion retailer. The objective of this study was to empirically analyze the performance of different pricing policies and their impact on revenues (and gross profit). Due to long production lead times, the retailer ordered inventory prior to the start of the selling season. While the inventory may be delivered to the stores over multiple periods during the selling season, little to no changes can be made to the order quantities during the selling season. Thus, the cost of acquiring and selling the items is, for the most part, a fixed or sunk cost; during the selling season the retailer can use price control to increase its sales revenues. The authors collected historical weekly data (sales and price) for 184 styles of women's apparel sold in 50 different store locations during the Spring 1993 selling season. (The Spring selling season extended from January 1993 through June 1993). The weekly sales data was fitted to a demand model; demand was assumed to be a function of the current selling price, seasonality, and a decaying age factor. This relatively simple demand model was found to perform quite well, with an average (over the 184 styles studied) R^2 value of 0.52.

The authors designed five different markdown pricing policies, and compared the actions taken (timing and magnitude of price changes) and profits achieved under these proposed strategies to the actions taken and profits achieved under the company's pricing policy. The five different markdown policies fall into two classes of strategies: (i) Full-information policies: These policies assume that full knowledge of the demand model and price sensitivity it known at the start of the horizon, and (ii) Adaptive policies: These policies assume that the retailer starts with initial estimates for the values of the demand function parameters. As demand is realized in each period, the retailer updates the demand function parameter estimates. Full-information policies, while unrealistic in practice, provide an upper bound on the potential benefits of dynamic pricing strategies.

We report here the results for one of the full-information policies ("FI-3") and one of the adaptive policies ("A-2"), and compare these results to those achieved under the company's pricing policy. Policy FI-3 assumes that the demand model is known with certainty. The policy restricts that the starting price and markdown timing must follow that adopted by the company's pricing policy.[7] However, the magnitude of the markdown can be optimally selected. Adaptive policy A-2 assumes that an initial estimate of the demand function parameters is given. This estimate is updated each period as demand is realized. An optimal price is then computed for each item sold. If (a) the item has been on the selling floor for at least three weeks and (b) the suggested optimal price is at least 20% lower than the current selling price then the new price is adopted. The results generated by these policies, as compared with those that the company achieved using its pricing mechanism are shown in Table 4.2.

Table 4.2. Comparison of retailer's actual pricing policy to proposed pricing policy FI-3 and pricing policy A-2

	Policy FI-3	Policy A-2	Company Policy
# markdowns	43	33	60
Avg. markdown %	22.4	25.3	25.8
Avg. markdown week	8.6	4.3	8.6
Revenue increase (%)	2.9	4.8	–

---

[7]   We select to report on only these two policies for the following reasons: Policy FI-1 is the most restrictive of all the full-information policies, as it requires the same initial price and markdown timing as the policy followed by the company. Policy A-2 is an example of a simple, model-driven, pricing policy that could be adopted in practice by a retailer.

The results reported in this table indicate that if the retailer had improved information regarding the true values of the demand function parameters, revenue could increase by 2.9% even if the retailer restricts himself to adopting the same initial price and markdown timing as the policy currently followed by the retailer. This revenue increase comes from two sources: (i) the average percentage markdown under policy FI-3 is 22.4%, somewhat lower than the 25.8% average markdown under the company's current pricing policy. (ii) Perhaps even more significant is the fact that the current pricing policy reduces the prices on 60 of the 184 styles, whereas policy FI-3 suggests price reductions on only 43 styles. However, a full-information policy is unrealistic in practice, as a retailer typically does not know the true value of the demand function parameters at the beginning of the selling season. For this reason we are interested in the performance of policy A-2, which is an example of a policy that can, in practice, be adopted by a retailer. Policy A-2 indicates a 4.8% increase in revenue over the existing company policy. This policy reduces the price on 33 of the 184 styles (as compared with the company's policy which reduced the price on 60 styles). While the average price reduction under policy A-2 is similar to that under the company's policy (average price markdown of 25.3% under policy A-2 versus 25.5% under the company's pricing policy), the timing of the markdown is quite different. Under policy A-2 the average markdown week is 4.3 versus week 8.6 under the company's pricing policy. The resulting increase in revenue seems to indicate that a company can achieve significant revenue increases by identifying, early on, which items should be marked down and then implementing these price reductions early in the selling season. Further, by reducing the price on 60 styles (rather than the 33 styles suggested by policy A-2) the company may have lost revenues from customers who were willing to purchase items at a higher price.

4.8 Concluding Remarks

This article reviews the most recent trends, in research and practice, in product pricing. Advances in research during the past 10 years and the availability of demand data from maturing information technology have created a favorable environment for a new industry of model-based price optimization to emerge. Many businesses have been eager to implement this new optimization tool as a means to improve revenue and/or profit.

It is yet too early to tell whether these systems will join the suite of standard information technology tools (such as an ERP system) adopted by businesses. Further, if the use of such systems becomes widespread, customers may develop strategic buying strategies that explicitly account for typical behavior of these price optimization algorithms. In addition, improved technology may help provide customers with extensive visibility into all candidate products available and their prices. (For example, in retail, online shopping bots help consumers identify all sources and prices for a particular product.) This technology could help the customer select the optimal product to purchase, given the total cost of purchase and the customers' product preferences. Both strategic buying strategies as well as tools to provide users with greater visibility into choices and costs, will modify traditional customer purchasing patterns. This may impact the effectiveness of the price optimization systems and may require the development of new models, such as those based on game theory.

References

Agosta, L. (2001) Sales logic: Price optimization software commands premium pricing in a soft software market, *IdeaByte*, Giga Information Group, September 3

Ahn, H., Gumus, M., Kaminsky, P. (2004) Pricing and Manufacturing Decisions when Demand is a Function of Prices in Multiple Periods, Working paper, Michigan Business School

Amihud, Y., Mendelson, H. (1983) Price Smoothing and Inventory, *Review of Economic Studies*, L, 87–98

Arjunji, R.V., Bass, F.M. (1996) A Model of Retail Promotion, Manuscript, Yale School of Management, New Haven, CT

Aviv, Y., Pazgal, A. (2002) Pricing of Short Life-Cycle Products through Active Learning, Working Paper, Olin School of Business

Aviv, Y., Pazgal, A. (2003) Optimal Pricing of Seasonal Products in the Presence of Forward-Looking Consumers, Working Paper, Olin School of Business

Bass, F.M. (1969) A New Product Growth Model for Consumer Durables, *Management Science* 15 (5): 215-226

Bernstein, F., Federgruen, A. (2001) Dynamic Inventory and Pricing Models for Competing Retailers, conditionally accepted to *Naval Research Logistics*

Ben-Akiva, M., Lerman, S. (1985) *Discrete Choice Analysis: Theory and Application to Travel Demand*, MIT Press, Cambridge, MA

Bitran, G.R., Caldentey, R., Mondschein, S. (1998) Coordinating Clearance Markdown Sales of Seasonal Products in Retail Chains, *Operations Research* 46 (5): 609-624

Bitran, G.R., Mondschein, S.V. (1997) Periodic Pricing of Seasonal Products in Retailing, *Management Science* 43 (1): 64-79

Blinder, A. (1982) Inventories and Sticky Prices: More on the Microfoundations of Macroeconomics, *American Economic Review* 72: 334–348

Boone, C. (2002) Casual Male improves gross margins and sell-throughs using ProfitLogic: An IDC customer case study, *IDC Research Report*

Cao, H., Gung, R., Lawrence, R., Jang, Y., Lin, G., Lu, Y (2002) Bid Winning Probability Estimation and Pricing Modeling, U.S. patent filed

Cassaigne, N. Papaioannou, V. (2000) Knowledge focused bid price setting process, *Proceedings of the 11th International Workshop on Database and Expert Systems Applications*, pp 841–845

Chan, L.M.A., Shen, Z.J., Simchi-Levi, D., Swann, J. (2004) Coordination of Pricing and Inventory Decisions: A Survey and Classification, Chapter 9 in: *Handbook of Quantitative Supply Chain Analysis: Modeling In The E-business Era*, Simchi-Levi, D., Wu, S.D., Shen, Z.J. (eds.), Kluwer Academic Publishers

Chan, L.M.A., Simchi-Levi, D., Swann, J. (2002) Dynamic Pricing Strategies for Manufacturing with Stochastic Demand and Discretionary Sales, Working Paper, Georgia Institute of Technology, Atlanta, GA

Chen, X., Simchi-Levi, D. (2002) Coordinating Inventory Control and Pricing Strategies with Random Demand and Fixed Ordering Cost, Working Paper, Massachusetts Institute of Technology, Cambridge, MA

Eliashberg, J.,Steinberg, R. (1991) Marketing-production joint decision making, in: Eliashberg, J., Lilien, J.D. (eds.), *Management Science in Marketing, Handbooks in Operations Research and Management Science*, Volume 5, North Holland, Amsterdam

Elmaghraby, W., Keskinocak, P. (2003) Dynamic Pricing in the Presence of Inventory Considerations: Research Overview, Current Practices, and Future Directions, *Management Science* 49 (10): 1287–1309

Federgruen, A., Heching, A. (1999) Combined pricing and inventory control under uncertainty, *Operations Research* 47 (3): 454–475

Feldman, J.M. (1990) Fares: To Raise or Not to Raise, *Air Transport World* 27 (6): 58–59

Feng, Y., Gallego, G. (1995) Optimal starting times for end-of-season sales and optimal stopping times for promotional fares, *Management Science* 41 (8): 1371–1391

Gallego, G., Huang, Y., Katircioglu, K., Leung, Y.T. (2000) When to share information in a simple supply chain, submitted for publication

Gallego, G., van Ryzin, G.J. (1994) Optimal Dynamic Pricing of Inventories with Stochastic Demand over Finite Horizons, *Management Science* 40 (8): 999–1020

Gallego, G., van Ryzin, G.J. (1997) A multi-product dynamic pricing problem and its applications to network yield management, *Operations Research*, 45 (1): 24–41

Gilbert, S.M. (2000) Coordination of Pricing and Multiple-period Production across Multiple Constant Priced Goods, *Management Science* 46 (12): 1602–1616

Girard, G. (2002) Price optimization is hot, but it's just the beginning, *AMR Research Report,* August 31

Hassin, R., Haviv, M. (2003) *To Queue or Not to Queue: Equilibrium Behavior in Queueing Systems* (International Series in Operations Research & Management Science, 59), Kluwer Academic Publishers

Heching, A., Gallego, G., van Ryzin, G. (2002) Markdown pricing: An empirical analysis of policies and revenue potential at one apparel retailer, *Journal of Pricing and Revenue Management* 1 (2): 139–160

Heching, A., Leung, Y.T., Levanoni, M., Parija, G. (2001) Optimal pricing and promotion decisions to maximize profit, presented at the INFORMS Marketing Science Conference, Wiesbaden, Germany

Ho, T.H., Tang, C.S., Bell, D.R. (1998) Rational Shopping Behavior and the Option Value of Variable Pricing, *Management Science* 44 (12): S145–S160

Johnson, C.A., Allen, L., Dash, A. (2001) Retail Revenue Management, *The Forrester Report*, December 2001

Krishna, A., Kopalle, P. (2003) Retailers' Optimal Ordering and Pricing Decisions in a Multi-brand Trade-Dealing Environment, Manuscript, University of Michigan Graduate School of Business

Lawrence, R.D. (2003) A Machine-Learning Approach to Optimal Bid Pricing, in: *Computational Modeling and Problem Solving in the Networked World: Interfaces in Computer Science and Operations Research*, Bhargava, H.K., Ye, N. (eds.), Kluwer Academic Publishers

Leung, Y.T., Levanoni, M., Hu, C.W., Ramaswamy, S. (2002) Better demand models through data clustering, presented at the INFORMS Marketing Science Conference, June 2002

Leung, Y.T., Ramaswamy, S. (1998) Price Markdown Assessment, proprietary client report, IBM Research Division, July 1998.

Maglaris, C., Zeevi, A. (2003) Pricing and design of differentiated services: Approximate analysis and structural insights, submitted for publication

Marn, M.V., Roegner, E.V., Zawada, C.C. (2003) The Power of Pricing, *McKinsey Quarterly*, Number 1

McWilliams, G. (2001) Lean Machine: How Dell Fine-Tunes its PC Pricing to Gain Edge in a Slow Market, *Wall Street Journal* (Eastern edition), June 8, p. A1

Merrick, A. (2001) Priced to move: Retailers attempt to get a leg up on markdowns with new software, *Wall Street Journal*, August 7

Monroe, K.B., Della Bitta, A.J. (1978) Models for pricing decisions, *Journal of Marketing Research* XV: 413–428

Neslin, S.A., Powell, S.G., Stone, L.S. (1995) The Effects of Retailer and Consumer Response on Optimal Manufacturer Advertising and Trade Promotion Strategies, *Management Science* 41 (5): 749–766

O'Neill, M., Daggupaty, V., Cauley, C. (2003) Max margin, The role of price optimization systems, *IDC Report*, Number 29704, Vol. 1, January 2003

Papaioannou, V., Cassaigne, N. (2000) A critical analysis of bid pricing models and support tool, *2000 IEEE International Conference on Systems, Man, and Cybernetics*, 3, 2098–2103

Petruzzi, N., Dada, M. (1999) Pricing and the Newsvendor Problem: A Review with Extensions, *Operations Research*, 47 (2): 183–194

Riley, J. and Zeckhauser, R. (1983) Optimal selling strategies: When to haggle, when to hold firm, *The Quarterly Journal of Economics*, 98 (2): 267–289.

Riley, J.G., Samuelson, W.F. (1981) Optimal Auctions, *American Economic Review* 71: 381–392

Scott, K. (2003) Driving demand profitability with pricing, *AMR Research Report*, January 2003.

Shankar, V., Bolton, R. (2004) An Empirical Analysis of Determinants of Retailer Pricing Strategy, *Marketing Science* 23 (1): 28–49

Silva-Risso, J.M., Bucklin, R.E., Morrison, D.G. (1999) A Decision Support System for Planning Manufacturers' Sales Promotion Calendars, *Marketing Science* 18 (3): 274–300

Smith, S.A., Achabal, D.D. (1998) Clearance Pricing & Inventory Policies for Retail Chains, *Management Science* 44 (3): 285–300

Subrahmanyan, S., Shoemaker, R. (1996) Developing optimal pricing and inventory policies for retailers who face uncertain demand, *Journal of Retailing* 72 (1): 7–30

Talluri, K.T., van Ryzin, G.J. (2000) Revenue Management Under a General Discrete Choice Model of Demand," *Management Science* 50 (1): 15–30

Tedechi, B. (2002) Scientifically priced retail goods, *The New York Times*, September 2

Tellis, G.J., Zufryden, F.S. (1995) Tackling the Retailer Decision Maze: Which Brands to Discount, How Much, When and Why?, *Marketing Science* 14 (3): 271–299

Tohamy, N., Johnson, C.A., Herbert, L. (2003) Price optimization adoption soars in 2004, *Forrester Research Brief*, December 5

Tsay, A.A., Nahmias, S., Agrawal, N. (1999) Modeling Supply Chain Contracts: A Review, in: *Quantitative Models for Supply Chain Management*, Tayur, S., Ganeshan, R., Magazine, M. (eds.), Kluwer Academic Publishers

Van Mieghan, J., Dada, M. (1999) Price vs. Production Postponement: Capacity and Competition, *Management Science* 45 (12): 1631–1649

Wolfstetter, E. (1999) *Topics in Microeconomics: Industrial organization, auctions, and incentives*, Cambridge University Press

Yano, C., Gilbert, S.M. (2003) Coordinated Pricing and Production / Procurement Decisions: A Review, in: *Managing Business Interfaces: Marketing, Engineering, and Manufacturing Perspectives*, Chakravarty, A.K., Eliashberg, J. (eds.), Kluwer Academic Publishers

Zhu, K., Thonemann, U.W. (2002) Coordination of Pricing and Inventory Control Across Products, Working Paper, The Hong Kong University of Science and Technology and University of Munster (under review)

5 Applications of Implosion in Manufacturing

Brenda Dietrich, Daniel Connors, Thomas Ervolina, J.P. Fasano, Robin Lougee-Heimer and Robert J. Wittrock

Manufacturing firms use a variety of software tools to improve the quality and speed with which strategic, tactical, and operational decisions are made. In this paper we first review the environment that lead to the development of "implosion technology" within IBM. We include a small illustrative example of the implosion problem, including a mathematical formulation. We then discuss the development and deployment of software tools for solving the implosion problem. A general mathematical formulation of the implosion problem is included in a separate section.

In the 1980's and early 1990's one of the most common computer-supported activities in manufacturing firms was material requirement planning. The APICS dictionary defined *material requirement planning (MRP)* as "a set of techniques that use bill of material, inventory data, and the master production schedule to calculate requirements for material." MRP was included in the larger process "Manufacturing Resource Planning (MRP II)," which was defined by APICS as "a method for the effective planning of all resources of a manufacturing company." In addition to MRP, MRP II provided other linked functions, including business planning, sales and operations, production planning, capacity requirements planning, and the execution support system for capacity and material. More commonly, MRP II (later ERP, for Enterprise Resource Planning) was used as a catch-all term to classify software tools that include MRP and some collection of planning, resource management, and accounting applications.

APICS also prescribed that "the master production schedule must take into account the forecast, the production plan, and other important considerations such as backlog, availability of material, availability of capacity, management policy, and goals." The master production scheduler needed to balance conflicting objectives and make trade-offs. Although it was acknowledged that computer software could greatly aid the master production scheduler, expert opinion was that human judgment would always be required (Vollman et al. 1992). Master production scheduling was also identified as one of the more critical but poorly addressed issues in manufacturing planning and control (Le Roy 1992). Manual "bottom-up re-planning" is defined by APICS as "... the process of using pegging data to solve material availability and/or problems. This process was accomplished by the planner (not the computer system), who evaluates the effects of possible solutions. Potential solutions include compressing lead time, cutting order quantity, substituting material, and changing the master schedule." Re-planning or re-scheduling became necessary as a result of delay, engineering specification changes, buyer order cancellations, previous scheduling errors due to lack of information, part shortages, or manufacturing quality problems (Park 1993). Lee and Billington (1991) state that in an internal Hewlett-Packard Company survey, managers reported that "incoming part availability and part delivery performance are the most important problems they face today." It is significant that as late as 1995 APICS advocated leaving the daunting re-planning task up to the planner and not to the computer.

With the availability of high-performance workstations, a new class of MRP-like tools emerged. These tools acted as a front-end to the MRP II system, allowing managers to examine all critical data, and to develop new business plans in response to changing conditions (Shepherd 1993). They typically extracted data from the host MRP system, and provided the planner with graphical applications for interactive planning and simulation of material planning, capacity planning, and master scheduling. Entire planning runs were re-calculated in minutes, allowing the planner to evaluate many alternative plans. Rapid MRP emulation allowed planners to quickly identify material and capacity shortages and inventory excesses associated with a production and procurement plan. These tools were used by planners in an iterative fashion to reconcile identified inventory and capacity shortages. Although these tools met many of the requirements for a "master production scheduling system" that were generated through surveys of European manufacturing companies in the late 1980's (Tierney, Higgins, and Brown 1991), they were not truly "schedulers" because they were only capable of evaluating a proposed schedule. These tools could not generate feasible master production schedules, let alone determine schedules that were both feasible and optimal with respect to economic factors.

Master production scheduling specifies an allocation over time of the available material and capacity resources to competing production activities. Material resources include both raw materials that are purchased from other firms and subassemblies that are produced internally. Capacity resources include manufacturing machines, tools, and operators. The key distinguishing factor between these two types of resources is that unused material remains available for later use, while unused capacity (e.g., machine idle time) does not. Capacity is measured in planning periods of a day, week, or month. Capacity feasibility means that, in each period, the total capacity used by the production schedule does not exceed the total capacity available (Silva 1992). Material feasibility means that the total amount of material consumed in any planning period does not exceed the amount of material available in that period.

Automated master production scheduling systems can result in greatly improved manufacturing efficiency and can significantly reduce the cycle time of the planning process (Arbon et al. 1994). A good master schedule is a prerequisite for any operational scheduling system. Resource allocation methods can also be used for detailed production scheduling, order-release scheduling, and final-assembly scheduling. Typical uses include determining daily production plans that make efficient use of on-hand inventory, re-planning in response to machine failures or quality problems, and customer ship-date quoting.

The Manufacturing Logistics group at the IBM Thomas J. Watson Research Center began studying the resource-allocation problem in 1989. This group provided modeling and analysis support and software to IBM manufacturing sites. In the late 1980's the PS/2 card plant in Austin experienced shortages of electronic components that were required for the production of several different PS/2 cards. IBM's world-wide material planning process required that the card plant commit an availability schedule for PS/2 cards within a few days of receiving a forecast of card requirements and an availability schedule for components. The card volume planners had no tools, other than simple spreadsheets, to aid in determining how to allocate the limited availability of the scarce components to cards, and on occasion had produced infeasible committed availability schedules. This lead to both lowered revenue, and increased costs resulting from excess inventory of other computer parts (disk drives, power supplies) that could not be used. Before discussing our solution, we present a small example of the resource allocation problem.

5.1 Manufacturing Resource-Allocation Models: A Small Example

The manufacturing resource-allocation problem is concerned with determining how much of each product to produce. When products share common resources, production quantities must be determined in a coordinated fashion, so that the total consumption of any resource does not exceed its availability. For example, suppose that there are three products, P1, P2, and P3 and two resources, R1 and R2. The usage rate of each resource for the production of each product is given in the following table.

Products		P1	P2	P3
Resources	R1	2	1	1
	R2	1	1	1

Assume that the demand for the products P1, P2 and P3, is 10, 5 and 15 units, respectively. Then, to fully satisfy these demands requires $(10\times2)+(5\times1)+(15\times1)=40$ units of R1 and $(10\times1)+(5\times1)+(15\times2)=45$ units of R2. Now suppose that only 30 units of R1 and 35 units of R2 are available. One might ask the question, "How much of P1, P2, and P3 can be produced?" There are many possible answers to this question. Two sample combinations of P1, P2 and P3, and the corresponding usage of resources R1 and R2 are given in the following table:

Product			Resource	Requirement
P1	P2	P3	R1	R2
10	0	10	30	30
0	5	15	20	35

In the first combination the entire supply of R1 is consumed, while in the second combination, the entire supply of R2 is consumed. Any combination of P1, P2, and P3 that does not consume more than the available quantity of each resource is said to be feasible with respect to the resource availability constraints. To describe all feasible combinations we require three decision variables:

x_{P1} = the number of units of P1 to be produced,

x_{P2} = the number of units of P2 to be produced, and

x_{P3} = the number of units of P3 to be produced.

Each unit of P1 requires 2 units of R1, each unit of P2 requires 1 unit of R1, each unit of P3 requires 1 unit of R1, and only 30 units of R1 are available, so any feasible solution must satisfy the linear inequality $2x_{P1}+x_{P2}+x_{P3}\leq30$. Similarly each unit of P1 requires 1 unit of R2, each unit of P2 requires 1 unit of R2, each unit of P3 requires 2 units of R2, and only 35 units of R2 are available, so any feasible solution must also satisfy $x_{P1} + x_{P2} + 2x_{P3} \leq 35$. Production quantities cannot be negative, so $x_{P1} \geq 0, x_{P2} \geq 0$, and $x_{P3} \geq 0$. To eliminate schedules that produce more of a product than is demanded we require that $x_{P1} \leq 10$, $x_{P2} \leq 5$, and $x_{P3} \leq 15$.

These eight inequalities define the set of feasible production schedules. That is, any combination of values of x_{P1}, x_{P2}, and x_{P3} that satisfies those eight constraints corresponds to a production combination of P1, P2, and P3 that can be made with the available quantities of R1 and R2. In general, a set of linear constraints can have zero, one, or many feasible solutions. The special structure of this resource-allocation problem eliminates the possibility of the constraints having no feasible solutions, since the "do nothing" solution obtained by setting all production quantities to zero trivially satisfies all of the constraints.

To select the best possible solution from among the set of feasible solutions, the criteria for determining "best" must be defined. Given a per-unit profit values for each of the production variables, the profit associated with a production schedule can be expressed as a linear combination of the decision variables. Suppose that the per-unit profit values of the three products are given by 6, 5, and 8, respectively. Then the profit associated with producing quantities x_{P1}, x_{P2}, and x_{P3} of products P1, P2, and P3, respectively, is given by $6x_{P1} + 5x_{P2} + 8x_{P3}$. Two feasible production schedules and their associated profit are given in the following table.

P1	P2	P3	Profit
10	0	10	$(6\times10)+(5\times0)+(8\times10)=140$
0	5	15	$(6\times0)+(5\times5)+(8\times15)=145$

The production plan which maximizes profit is found by solving the following linear programming problem:

$$\text{MAX } 6x_{P1} + 5x_{P2} + 8x_{P3}$$
$$\text{subject to } 2x_{P1} + x_{P2} + x_{P3} \leq 30$$
$$x_{P1} + x_{P2} + 2x_{P3} \leq 35$$

$$x_{P1} \geq 0 \qquad x_{P3} \geq 0 \qquad x_{P2} \leq 5$$
$$x_{P2} \geq 0 \qquad x_{P1} \leq 10 \qquad x_{P3} \leq 15$$

Linear programming is a well studied mathematics discipline. Solution algorithms were developed in the 1940's and small linear programming problems were being solved by computer implementations as early as 1953. The following five decades have seen rapid advances in linear programming software. In fact, through a combination of dramatically improved computer hardware and advances in solution algorithms the size of linear programs (expressed as number of constraints) considered reasonable to solve has grown exponentially since 1950 (Orden 1993; Nemhauser 1994; Bixby 2000). The optimal solution to this small example problem is

P1	P2	P3	Profit
$6\frac{2}{3}$	5	$11\frac{2}{3}$	$158\frac{1}{3}$

The schedule $x_{P1} = 6$, $x_{P2} = 5$, $x_{P3} = 12$, which has profit of 157 is the optimal integer solution.

The difference between resource-allocation-based planning and traditional MRP and CRP can be understood in terms of the inputs, assumptions, and outputs of these two methods. Both MRP and CRP consider the top-level demand (MPS) to be fixed input data and assume infinite material and capacity availability. MRP calculates required supply quantities and CRP calculates required capacity levels, and both generate recommendations for changes to supply orders. In contrast, resource-allocation models take the material and capacity availability to be known, finite input data, and treat the top-level demand as a desirable but not necessarily attainable target. These models calculate modifications to the MPS that ensure feasibility and optimize specific economic criteria. In addition, resource-allocation models can be extended to consider factors that cannot be represented with traditional MRP/CRP methods, such as allocation of production to customers or demand classes, use of substitute material, and allocation of production to alternative capacity sources.

In resource-allocation-based planning, the production quantities for each product in each period are decision variables. Constraints on the decision variables are determined from the capacity-availability limits, the bill-of-capacity structure, the material-availability limits, the bill-of-material structure, and the original demand schedule. These constraints limit the values that can be simultaneously taken by the decision variables. Profit or serviceability maximization is used as an objective function, and the problem is solved through the use of heuristics or linear programming algorithms. Extensions, such as allocation of production to specific customers, require additional decision variables, such as the quantity of each product shipped to each customer in each period, and constraints. Further discussion of LP based modeling in an MRP environment including a more complex single-period example can be found in (Bahl, Taj and Corcorans 1991). A general formulation of a linear program for the implosion problem can be found in Section 5.4.

5.2 An Implosion Heuristic

The implosion problem can be represented and, for moderate size problems, solved as a linear program (see Section 5.4), using standard commercial software such as IBM's Optimization Solutions and Library or ILOG's CPLEX, or open source software included in the COIN-OR repository (www.coin-or.org). However, during the early years of the implosion project, the computational speed of the available hardware, coupled with limitations of the available software, restricted the practicality of using of linear programming in implosion applications. Therefore, an alternate solution method, known as the "implosion heuristic" was developed. The implosion heuristic is intended to quickly produce feasible, near optimal solutions. It takes as input standard MRP data: bills of material, supply of parts and capacity, and demand; it and produces as output a feasible production plan and shipment schedule. It can also provide reports on backlog, unused capacity, and stock levels of parts. The heuristic is further described in Dietrich and Wittrock (1996).

The idea behind the implosion heuristic is quite simple, and is based on the fact that given a finite supply of resources, for a single demand element (that is, a part, time period, quantity triple p,t,N) there is a maximum quantity of p that can be completed in time period t. Specifically, assume that p has no substitutes specified in its bill of material, and that all production must be done exactly according to the offsets specified in the BOM. Then we can determine whether n units of the part can be produced in t by simply exploding a single external demand against the available supply and capacity using standard MRP explosion. If the explosion does not produce any net requirements, then the quantity n can be produced; if the explosion produces any net requirements, it cannot. Further assume that the minimum production quantity of p is a single unit, and note that if n units of p can be produced in t, then each of the smaller quantities $n-1, n-2,...,1$ of p could also be produced in t, while if n units of p cannot be produced in t, then no larger quantity $n+1, n+2, ...,N$ can be produced in t. Thus, under these limiting assumptions, one can determine the maximum quantity of part p that can be produced in period t by using a binary search that that calls an MRP explosion.. By considering the demands in some specified order, and for each demand determining the maximum quantity of the demand that can be met in the prescribed time period and subtracting the resources used to meet this demand from the supply, one can produce a feasible production plan and corresponding shipment plan. Computational efficiency and solution quality can be achieved through extensions of this simple approach.

To understand the computational performance of the above approach, first note that a single demand explosion, or test for feasibility, involves considering only those resources that are in the bill-of resources tree for the part p. The explosion process requires many multiplication and addition operations, but the number is bounded by the product of the number of items in the bill of resources tree and the number of time periods. The explosion can stop with the demand quantity found infeasible once a net requirement for a capacity or a raw material is detected. The number of feasibility tests required is also bounded. If remaining demand is simply discarded if it can not be met by just in time production, then the number of feasibility tests is at most the sum of the quantities $log (N)$ over all demand triples (p,t,N). The number of explosions can be reduced, in the case where demand quantities are large, by imposing a minimum production quantity and a minimum production increment. Tighter bounds on the expected number of explosions required can also be obtained by noting that in the vast majority of searches for the maximum feasible production quantity, either the entire quantity can be produced, or no units of the part can be produced. Thus most of the production quantities are determined by at most two explosions.

When all resources are readily available, the above approach produces a production plan that meets all demands through just in time production. If there are shortages of some resources, the production of parts that require those resources will be limited by the availability of the resources. If a scarce resource is used in only one part, the production of that part will be reduced so that it does not exceed the availability of that resource. For capacity resources, this reduction is on a period by period basis; for material resources, where supply from an earlier period can be used in later periods, the total consumption through any period cannot exceed the total availability through that period. In the more typical case, where a scarce resource is used by multiple parts, the order in which the demands are considered is the primary factor determining the allocation of the scarce resource. The demands that are considered first will be met and will consume the resources, leaving none for the demands that are considered later. By ordering the demands according to business objectives, high quality solutions can be obtained. If equitable solutions, which evenly share scarce resources, are desired, each demand can be broken into several smaller demands, and these demands interleaved in the order.

This basic heuristic approach can be extended to deal with substitutions, various forms of build-ahead, and some reallocation of stock. In all of the heuristic extensions, the "no-backtracking" principle has been maintained. That is, once a production quantity has been determined to be feasible, that quantity is never later reduced. Adhering to this design principle has allowed us to customize the heuristic to address a number of complex scenarios without compromising execution speed. One particularly important form of customization addresses business rules related to backlog. If a demand cannot be completely met in the requested period, the remaining quantity can be ignored (appropriate for the case where customers will substitute a competitor's product), it can be added to demand for that part in the following period, or it can be used to create a new, high priority demand in the following period.

5.3 IBM usage of implosion

Implosion technology has been used throughout IBM's supply chain for a variety of applications that required a rapid assessment of capability to respond to changes in demand, supply, or capacity. The first systematic use of implosion was in the Austin, Texas, card assembly plant, which produced circuit cards for IBM personal computers. The Enterprise Shortfall Implosion Tool (ESIT) was jointly developed by IBM's Research division and its Corporate Logistics group, and released to the Austin card plant in April 1990. The tool was written in PL/1, used manufacturing data from relational databases, and ran on MVS. It considered only a single level build structure in which each part number was either a raw material (component) or a finished product (card). ESIT helped production planners in the Austin plant close the monthly planning cycle by providing feasible commitments for card production volumes. These card production plans needed to account for shortages in key components that were used by multiple card types, as well as for long lead times on modules and limits on card tester capacity. Use of the tool simplified and shortened the Austin planning process, and provided the PC division with greater understanding of the impact of component shortages on PC production. Its use was linked to overall improvements in PC component inventory levels and faster responses by the PC division to shifts in demand. In addition, outside of the official planning cycle, the tool was used for a variety of other purposes such as determining the impact of expediting a shipment or adding additional supplier capacity. ESIT included an LP-based solver, a heuristic solver, and an emulation of MRP explosion. The ESIT heuristic allocated components to cards based on demand date and priority, using a critical ratio heuristic similar to the one described in Luss 1986. The MRP emulation was used by the planners for off-cycle calculations of part and capacity requirements. Following the successful deployment of implosion in Austin, box plants, including the PC plant in Raleigh, North Carolina, and the AS/400 plant in Rochester, Minnesota, requested that the capability be provided to them.

Release 2 of ESIT, which included multi-level implosion and explosion, was provided to the Rochester plant in 1992. ESIT was used there for various production planning processes through 1994, when it was replaced by a workstation based version of the implosion tool. The critical ratio heuristic used in the first release of ESIT did not extend to multi-level build structures, especially those involving independent demand and supply of subassemblies, so a version of the heuristic described above was implemented. Various enhancements, such as inclusion of substitute parts and capacity, multiple demands for a single product with different economic factors or priorities, and stocking rules related to engineering changes in a bill of materials, were also added to ESIT. Although the tool was well integrated into planning processes, and provided significant value, adding new capability and using the base code for new applications became increasingly difficult. In 1992 development of a workstation version of the implosion tool, which would take advantage of developments in programming languages and environments, and exploit the performance of RISC workstations, was begun.

The Workstation Implosion Tool (WIT) was implemented in C on and RS6000 using object oriented design principles. WIT replaced a significant portion of the ESIT code in the Rochester implementation in 1993, but the existing interfaces with Rochester systems and reporting processes were maintained. Since 1994 the WIT tool has been used throughout IBM for a number of applications. An integration and application development framework was developed in 1994-1995. This framework was called PRM, for Production Resource Manager (also for "MRP spelled backwards"). See Dietrich et al (1999) for details on the PRM framework. The base WIT tool has been maintained and enhanced by the Research division, while the majority of the applications have been written by programming teams affiliated with other IBM divisions. Applications of WIT were a key component in IBM's supply chain reengineering program, which brought the 60-plus day supply/demand closure cycle down to 20 days.

In 1993-1994 WIT was used by the Personal Computer Division for box-level production and material planning in conjunction with a vendor's fast MRP emulation tool, called ESAT. The combined system, called WITerate, drove a number of additional WIT requirements, including extension of the heuristic to account for limited build ahead due to constrained capacity, allocation of substitute raw materials, and creation of a list of prioritized shortages that should be the focus of the material planners. The most extensive use of WIT within the Personal Computer division was at the Greenock, Scotland, and plant. Greenock developed the initial process and implosion application (inputs, reports, and run control was through VM) in 1996. In 1998, the Greenock application was deployed to Personal Computer Division sites at Raleigh and Guadalajara, Mexico.

In 1994 the implosion team also began to use the WIT tool to model semiconductor production. Speed sort and other separation processes in which one part number (e.g., wafer) is divided into two or more part numbers (e.g., fast chips and slow chips), were modeled through the use of negative usage rates on bill of material arcs. IBM's Microelectronics Division (MD) engaged the WIT team for several modeling exercises during this period. Over the next several years applications of WIT were developed for MD. One provided basic multi-site implosion to compute a Supply Commit for MD top level parts (those parts being provided to other divisions). This supported weekly execution planning by providing an "available to promise" capability (long before the "ATP" term became in vogue). The team also developed ASCOPA, which stands for Alternate, Substitute, Co-Products Advanced Planning System. The application provided an optimized MRP (explode) function that minimized the net requirements at the lowest component level by using implosion-based allocation of resources to reflect speed-sorting (co-product) and down-binning (substitution) structure associated with semi-conductor production.

In 1995 the Rochester box plant also began development of a WIT based tool that supported online customer order scheduling and manufacturing line scheduling. This application, called COMBAT, used real-time information on plant capacity, component supply, and the availability of critical resources. It was part of a division-wide deal creation/commit/completion process developed as part of IBM's supply chain reengineering initiative. COMBAT was integrated with the production and inventory control systems in three IBM divisions through the exploitation of client/server architecture. It was capable of using data from MAPICS, CIIM and CIMAPPS. The COMBAT application was developed using the PRM Framework by a team in Rochester, with assistance, and WIT and PRM enhancements provided by the Research team. In 1996, IBM undertook a massive re-engineering effort to redesign and integrate the manufacturing, planning, fulfillment, scheduling, and logistics functions into single Integrated Supply Chain.   The success of the Implosion models described above at individual plants and the experience of Closed-Loop planning processes provided motivation to design a Closed-Loop, constraint based planning process for the Integrated Supply Chain.   This sub-process of the overall Integrated Supply Chain was called the Enterprise Capability Assessment (ECA) Process.

The challenge of the ECA process was to combine the BOM structures of multiple plants into a single extended product structure representation of the full IBM Supply Chain. Based on this extended product structure, the ECA process called for centralized MRP and Implosion systems that could enable a single, closed-loop process for the entire supply chain.

For ECA, a new WIT application was developed called Supply Capability Engine (SCE). SCE required enhancements to WIT but also provided extensions to the core WIT model through customized application development. The key extensions that SCE provided to the implosion problem are: multi-site capability, explicit models for Configure to Order product, supply aggregation, Focused Shortage Schedule (FSS), and a Parts Conditioning File. For multi-site, an Interplant Relationship structure was added that defines a parent-child dependency between parts at different sites. The Parts Condition File enables users to list those raw materials whose supplies were constrained; all other raw materials were assumed to be available as needed. This reduced the data complexity of the problem and focused users on the real constraints. The FSS (a function of WIT) is a diagnostic utility that determines what critical resources were needed to improve the implosion solution for a subset (the Focus Set) of the end-items. The ECA process relied on the FSS to provide an updated component requirements schedule that could be passed on to suppliers. Thus, the ECA process specified two implosion runs: the first was based on initial supplier commits, and the second (and final) was based on the Supplier's commit to the FSS. By getting the suppliers to commit to a smaller set of critical parts (usually with a requested quantity between the original request and the original commit), the second implosion resulted in an improved Implosion answer overall.

In 1998, IBM replaced the ECA process with a Local Execution process. The central implosion of ECA caused a tight dependency across the supply chain which required very accurate representation of constraint data. One incorrect constraint could ripple through the entire structure. In response, IBM deployed a localized process that required each site to run its own, independent, five day closed loop cycle. The distributed approach of local execution removed the tight dependency and allowed each site to have better control of its implosion model. However, it also introduced a cascade effect into IBM's vertically integrated supply chain. The tier 1 box plant must wait for a commit from its tier 2 before it can commit end-product, while the tier 2 must wait for a commit from tier 3 before it can commit. So, while Local Execution enabled each site to optimize locally, the overall effect on the enterprise was sub-optimal.

The ECA and Local Execution processes represent extremes of (resp.) centralized and localized control. The experience at IBM of these two extremes led to a hybrid where the IBM supply chain was partitioned into three Value Chains: Complex Configured products (mainframe and server manufacturing), High Volume Easily Configured products (mostly PC's and printers), and Technology Products (semi-conductor and hard disk). Each Value Chain is structured as a centralized enterprise model, with implosion models tailored to each specific environment.

In 2001, IBM began using implosion models for ATP (Available to Promise) Generation within the Complex Configured value chain. The ATP Generation problem is to create a feasible production plan that can be used to schedule (or promise) orders against. The production plan is generated by an implosion model and it determines an optimal supply plan at the ATP push-pull boundary of the product structure. See Chen, Zhao, and Ball (2002) for more details on the ATP push-pull boundary. In a Build-to-Plan environment, the ATP push-pull boundary is at the end-product level, so the ATP Generation problem is similar to the classic implosion problem which seeks to create a feasible, optimal Master Production Schedule.

In a Configure-to-Order (CTO) environment, the ATP push-pull boundary is at the "feature" level. Features are the top-level component of the BOM. Each system shipped is a potentially unique configuration of Features selected from the "configuration menu" of the particular Machine Type being ordered. The ATP Schedule is specified in terms of overall Capacity at the Family level as well as availability of Features. An Allocated ATP Schedule is one where the Features are pre-allocated to Product Families. The BOM structure for CTO treats the usage rate of Features to Product Families as forecasted, flexible "attach rates". The implosion model for Allocated ATP in a CTO environment seeks to compute the maximum feasible supply of end-products at average configuration. In this case, the average configuration of the product is one where every feature is present on the product at its forecasted attach rate. Since there are many features and some of them are planned at very low attach rates, the implosion model allows the attach rates in the implosion solution to differ from their exact forecasted attach rates. Controlling exactly how the attach rates might diverge depends on what physical conditions are specified for the particular Feature as it applies to the Product.

SCE (with updated functionality), currently provides Allocated ATP generation for the Complex Configured (CC) value chain. The CC value chain is characterized by a CTO structure where the configuration menu has complex rules for what features are mutually selectable on any given order. The complex CTO model has driven the need for more custom constraints and algorithmic techniques to be developed in SCE. One other improvement to the overall SCE implosion model is the ability to relax the effect of a tightly integrated, centralized enterprise model. This is done through a combination of flexible CTO constraints and solving the implosion in stages.

Available to Sell (ATS) is a WIT based implosion application that provides optimization as well as squared set analysis for the consumption of excess inventory (on hand, liability, short-term overage), by finding saleable items that consume the excess while minimizing additional purchase. It's embedded into other planning tools and is available through a web interface. In the standard ATS implosion model, there is no explicit demand statement so every product is assumed to have infinite demand.   As is usually the case, the standard implosion model must be customized for use in a production environment.   For ATS, this means that a combination of WIT heuristic and WIT linear programming solvers are used to create custom implosion logic.

WIT has also been used to support IBM businesses other than manufacturing. IBM has a long history of product end of life management and recycling activities including a product take back option to corporate customers involved in a new sale or lease. IBM Global Finance's Asset Disposition and Support Services (GARS) for enterprise customers offers a range of end of life management services for IBM and non-IBM equipment alike. GARS uses an implosion application which takes a supply of returned machines and determines if they should be sold as used machines or disassembled to meet demand for used parts. The GARS implosion models make heavy use of co-product structures. From a single component (a complete system), the disassembly operation produces a large number of components. In some cases, the harvested components can have very high value and can be re-conditioned into usable parts again.

5.4   Mathematical Formulation of the Implosion LLP

The general implosion problem can be modeled as a linear program. The model presented here distinguishes between material resources (parts) and capacity resources. Multi-level bills of material are considered, although they are represented as a cascade of single-level bills of material. In addition, the model assumes that each part can have one or more demands placed on it, and that these demands can have different economic factors such as profit, revenue, or backlog penalties. First we discuss input data describing the manufacturing process. Then we define the decision variables and formulate the resource availability constraints. We then develop additional notation for economic factors and formulate a very general objective function. Additional modeling features such as substitute parts, substitute capacities, and decision variable bounds, can be added to the formulation, but are not discussed here. For details see WIT User's Guide and Reference (2004).

In most cases the resource-allocation problem requires consideration of multiple time periods. Production quantities, shipment quantities, stock levels, and scrap quantities for several periods must be considered. These periods need not be of equal length. The external availability of part resources for each period is known (or at least predictable) based on existing supplier orders, procurement lead-times, and on-hand inventory. Capacity availability for each period is also assumed known, or to be computable from current machine status, staffing plans, scheduled maintenance, and failure and repair statistics. The demand quantities for each demand in each period are estimated based on backlog, customer orders, and forecasts. Bills of material, bills of capacity, bounds, substitutability, and economic factors may vary from period to period.

Data:

J = set of parts (raw material, subassemblies, and end products)

T = set of time periods

R = set of capacity resources

$D =$ set of demands

$v_{j,0}$ = initial stock of part j

$e_{j,t}$ = net external supply of part j in period t

$a_{i,j,t,\tau}$ = quantity of part i required in period t per unit of part j produced (completed) in period τ

$c_{r,t}$ = quantity of capacity r available in period t

$g_{r,j,t,\tau}$ = quantity of capacity r required in period t per unit of part j produced (completed) in period τ

$p(d)$ = part for demand $d \in D$

$q_{d,t}$ = quantity of demand d in period t

$b_{d,0}$ = initial backlog for demand d.

The values $a_{i,j,t,\tau}$ define the bill-of-material structure and the values $g_{r,j,t,\tau}$ define the bill-of-capacity structure.

Decision Variables:

$x_{j,t}$ = quantity of part j produced (completed) in period t

$s_{d,t}$ = quantity of demand d filled in period t

$b_{d,t}$ = backlog of demand d at end of period t

$v_{j,t}$ = stock of part j at the end of period t

$\hat{v}_{j,t}$ = quantity of part j scrapped at the end of period t

$u_{r,t}$ = quantity of resource j unused at the end of period t

As indicated by their definitions, all of the model variables are required to be non-negative.

$$x_{j,t} \geq 0, \qquad\qquad \forall \ j \in J, \ t \in T$$
$$\hat{v}_{j,t} \geq 0, \qquad\qquad \forall \ j \in J, \ t \in T$$
$$v_{j,t} \geq 0, \qquad\qquad \forall \ j \in J, \ t \in T$$
$$s_{d,t} \geq 0, \qquad\qquad \forall \ d \in D, \ t \in T$$
$$b_{d,t} \geq 0, \qquad\qquad \forall \ d \in D, \ t \in T$$
$$u_{r,t} \geq 0, \qquad\qquad \forall \ r \in R, \ t \in T$$

Backlog is defined to be cumulative quantity demanded minus cumulative shipments. Since $\sum_{\tau \leq t} q_\tau = q_{d,t} + \sum_{\tau \leq t-1} q_{d,\tau}$ and $\sum_{\tau \leq t} s_{d,t} = s_{d,t} + \sum_{\tau \leq t-1} s_{d,\tau}$ we have $b_{d,t} - b_{d,t-1} = q_{d,t} - s_{d,t}$, or, $b_{d,t} - b_{d,t-1} + s_{d,t} = q_{d,t}$, $\forall d,t$

The capacity availability constraints say that for each capacity, the unused availability in each period is equal to the original availability minus the total amount of capacity used in that period in the production of parts. Since the leftover capacity $u_{r,t}$ is required to be non-negative, this means that the production schedule cannot use more capacity than is available in any period

$$\sum_{j \in J, \tau \in T} g_{r,j,t,\tau} \, x_{j,\tau} + u_{r,t} = c_{r,t} \qquad\qquad \forall \ r, t$$

The most complex equality is the material balance constraint. Unused material resource can be carried over from one period for use in a later period. Recalling that $x_{j,t}$ is the quantity of part j completed in period t and applying the time-phased usage rates, the material balance constraint takes the form:

$$\sum_{d \in D: p(d)=j} s_{d,t} + \sum_{k \in J, \tau \in T} a_{j,k,t,\tau} \, x_{k,\tau} + \hat{v}_{j,t}$$
$$+ v_{j,t} - x_{j,t} - v_{j,t-1} = e_{j,t} \qquad\qquad \forall \ j, t$$

Since all decision variables are required to be non-negative, this constraint says that the total consumption of a part (for meeting demand, for production of other parts, or as carry-over stock) cannot exceed the total supply of the part (from external supply, previous period stock, or production), with the balance being scrap.

These constraints define the feasible production schedules. Additional manufacturing considerations such as yield and fallout can easily be incorporated into the usage coefficients in the inventory and capacity balance constraints.

Note that the number of variables is given by $3 \times [J] \times [T] + 2 \times [D] \times [T] + [R] \times [T]$

Aside from the non-negativity constraints on all of the model variables, the model has $([J] + [D] + [R]) \times [T]$ constraints. For realistic manufacturing problems the LP models become quite large. For example in a problem with 1000 parts, 500 demands, and 50 resources the one period resource-allocation problem has 4050 variables and 1550 complex constraints and the 25 period problem has 101,250 variables and 38,750 complex constraints. For 20,000 parts, 1000 demands, and 100 resources, the 1 period problem has 62,100 variables and 21,100 complex constraints, and the 25 period problem has over a million variables and half a million constraints.

Additional data is required to express the objective function.

$S_{r,t}$ = scrapping cost per unit of part j in period t

$H_{j,t}$ = holding cost per unit of part j in period t

$M_{j,t}$ = manufacturing cost per unit of part j completed in period t

$R_{d,t}$ = revenue per unit of demand d shipped in period t

$P_{d,t}$ = penalty per unit backlog of demand d in period t

$Q_{r,t}$ = penalty per unit excess of capacity resource r in period t

An objective function that maximizes profit is given by:

$$\text{MAX} \sum_{t \in T} \left(\sum_{d \in D} R_{d,t}\, s_{d,t} - \left(\sum_{j \in J} H_{j,t}\, v_{j,t} + \sum_{j \in J} S_{j,t}\, \hat{v}_{j,t} \right. \right.$$

$$\left. \left. + \sum_{j \in J} M_{j,t} + \sum_{d \in D} P_{d,t}\, b_{d,t} + \sum_{r \in R} Q_{r,t}\, u_{r,t} \right) \right)$$

This formulation permits the case of negative usage quantities, where producing part j creates rather than consumes part i. This permits modeling the production of co-products and by-products. The formulation can be augmented to include minimum and maximum release quantities for each part, minimum shipments per demand and minimum and maximum stock quantities for each part.

References

Arbon, R., Mally, G., Osborne, T., Riethmeier, P., Tharrett, R. (1994) Auto-MPS: An Automated Master Production Scheduling System for Large Volume Manufacturing, *Proceedings of the Conference on Artificial Intelligence Applications*, 26–32

Bahl, H., Taj, S., Corcorans, W. (1991) A Linear-Programming Model Formulation for Optimal Product-mix Decisions in Material-Requirements-Planning Environments, *International Journal of Production Research* 29, 1025–1034

Bixby, R. (1994) Progress in Linear Programming, *ORSA Journal on Computing* 6, 15–22

Burcher, P. (1991) Closing the Loop in Manufacturing Resource Planning Systems, *Control*, August-September, 35–39

Chen, C., Zhao, Z., Ball, M.O. (2002) A Model for Batch Advanced Available-to-Promise, *Production and Operations Management* 11 (4): 424–440

Clode, D. (1993) A Survey of U.K. Manufacturing Control Over the Past Ten Years, *Production and Inventory Management Journal* 34 (2): 53–56

Dietrich, B. and Escudero, L. (1990) A Single-level Implosion Tool with Replacement Parts and Effectivity Dates, *Proceedings of the 1990 IBM Manufacturing Productivity Symposium,* Thornwood, New York, October 1990

Dietrich, W., Dietrich, B., Ervolina, T., Fasano, J., Lougee-Heimer, R., Poole, E., Tang, J., Wang, R., Wittrock, R., Wong, D., Chin, G. (1999) Production resource manager (PRM) framework, Chapter in Volume 3: *Domain-Specific Application Frameworks: Frameworks Experience by Industry,* Fayad, M., Johnson, R. (eds.), John Wiley and Sons

Dietrich, B.L., Wittrock, R.J. (1996) Allocation Method for Generating a Production Schedule, Patent #5,548,518, 8-20-96

Dietrich, B.L., Wittrock, R.J. (1997) Optimization of Manufacturing Resource Planning, Patent #5,630,070, 5-13-97

IBM (1990) *Optimization Subroutine Library Guide and Reference*, Release 2, SC23-0519-02

IBM (2004) *Watson Implosion Technology, User's Guide and Reference*, Release 6.0

Lechner, L., (1990) Critical Commodity Allocator (Implosion Prototype), *Proceedings of the 1990 IBM Manufacturing Productivity Symposium,* Thornwood, New York, October 1990

Lee, H., Billington, C. (1993) Material Management in Decentralized Supply Chains, *Operations Research* 41, 835–847

Le Roy, P. (1992) Master Production Scheduling and Management of Constraints: A Case Study, *Proceedings of the Eighth CIM-Europe Annual Conference*, Birmingham, UK, May 1992

Lopez, D. and Haughton, D. (1993) World-class Manufacturing in Northwest Electronics Companies: Distressing Results for Technology and MPC Applications, *Production and Inventory Management Journal* 34 (4): 56–60

Luss, H., Smith, D. (1986) Resource Allocation Among Competing Activities: A Lexicographical Minimax Approach, *Operations Research Letter* 6, 227–231

Lustig, I., Marsten, R., Shannon, D., Interior Point Methods for Linear Programming: Computational State of the Art, *ORSA Journal on Computing* 6, 1–14

Nam, S., Logendran, R. (1992) Aggregate Production Planning – A Survey of Models and Methodologies, *European Journal of Operational Research* 61, 255–272

Nemhauser, G. (1994) The Age of Optimization: Solving Large-Scale Real-World Problems, *Operations Research* 42, 5–13

Orden, A. (1993) LP from the '40s to the '90s, *Interfaces* 25, 2–12

Shepherd, J. (1993) The Next Wave in Planning Tools: The Decision Server, *APICS-The Performance Advantage*, May, 50–53

Silva, D. (1992) Capacity Management: Get the Level of Detail Right, *APICS International Conference Proceedings*, October 1992

Sounderpandian, J., Balashanmugam, B. (1991) Multiproduct, Multifacility Scheduling Using the Transportation Model: A Case Study, *Production and Inventory Management Journal* 32 (4): 69–73

Tierney, K., Higgins, P., Brown, J. (1991) User Requirements for Development of a Master Production Scheduling System, *Proceedings of the Seventh CIM-Europe Annual Conference*, Torino, Italy, May 1991, 211–220

Vollman, T.E., Berry, W.L., Whybark, D.C. (1992) *Manufacturing Planning and Control Systems*, 3rd Edition, McGraw-Hill College

Wilson, D.G., Rudin, B.D. (1992) Introduction to the IBM Optimization Subroutine Library, *IBM Systems Journal* 31 (1): 4–10

6 Strategic Sourcing and Procurement

Robert Guttman, Jayant Kalagnanam, Rakesh Mohan, and Moninder Singh

6.1 Background

Recent developments in IT have focused on providing a platform that facilitates and streamlines the activities of the purchasing department within an enterprise. Cost savings that are realized in these activities have a direct impact on the bottom line of an organization and the growing number of testimonials about excellent ROI has prompted companies to consider sourcing and procurement as the next large IT investment. The purpose of this chapter is to provide an overview of the various functions in sourcing and procurement and the techniques that are required to enable a large enterprise to rationalize their sourcing and procurement function.

In an enterprise there are two kinds of procurement:

1. **Direct Procurement:** relates to the commodities and parts that are used in the products/services produced and sold by a company. For example, a computer manufacturer would procure chips, memory, hard drive, monitors etc for assembling computers as opposed to a chocolate manufacturer that needs to buy sugar, milk, cocoa etc for manufacturing chocolates. This implies that the specification of the commodities changes quite dramatically across industry verticals and one needs to develop a repository of features differentiated by industry. In addition, sourcing of direct commodities is directly influenced by the design and manufacturing functions within the organization. For example, the screws that are used in the design of a pump need to consider the standard sizes that are available and the supply pool available from within the procurement organization. Such a close interaction between design and procurement makes the sourcing exercise more complex and might also require that the specifics of the commodity reflect the design context within which it is to be used.

2. **Indirect Procurement:** relates to commodities that are used by any enterprise for its day to day operations but which are not part of its main production. For example office stationary, office furniture, are typical examples of indirect procurement. These commodities are largely common across industries and there is large opportunity to standardize the specification of such commodities. The sourcing process can be increasingly automated and streamlined. A fundamental difference of indirect procurement is that the commodities are being requisitioned by a very large number of people across the enterprise and this requires that the procurement functions (such as requisitioning, supplier catalogs and payments) are deployed enterprise wide.

In this chapter, we focus on identifying the core functions and the techniques that underlie the sourcing and procurement functions across all industries for both direct and indirect procurement. We will leave out any discussion that relates to the development of industry specific knowledge based systems required to support sourcing and procurement such as industry specific parts catalogs or feature repositories.

6.2 Overview

Figure 6.1 provides an overview of the main activities in sourcing and procurement.

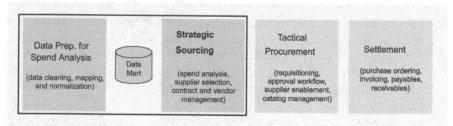

Fig. 6.1. Overview of a Purchasing Department Activities in a Corporation

1. **Spend Analysis:** The focus of this activity is to develop an aggregate view of the procurement spend across the organization using the transaction data. The aggregate spend by commodity, supplier plant etc provides a basis for identifying cost saving strategies. A typical example is to find commodity classes or plants where reducing suppliers and increasing volume to a small number of (preferred) suppliers might allow for better price negotiations. Another piece of analysis is

to track the performance of each supplier based on past behavior. This is a strategic activity.

2. **Sourcing**: One of the fundamental aspects of sourcing is supplier selection (for a commodity class identified by spend analysis) using one of many negotiation techniques (such as RFx, auctions etc). Once the suppliers are selected the relationship with the selected suppliers is then managed through the negotiated contracts. This step operationalizes the strategy developed by spend analysis.

3. **Procurement**: is a tactical activity where purchasing is (ideally) performed within the umbrella of existing contracts. Typical purchasing within an enterprise starts with a requisition that is approved and purchased from within catalogs of selected suppliers. An additional activity that is supported at this level is the enablement of (new) supplier catalogs and the management of these catalogs.

4. **Settlement:** is the follow through activity where the purchase is ordered, invoiced etc. This is the routine of bookkeeping of each purchase.

In this chapter, we examine each of these four activities and discuss the techniques that we have developed from a decision support perspective. The primary focus of this chapter is on strategic sourcing since this is the area where analytics is primarily used. Procurement and Settlement typically are focused around managing transactions and book keeping and we will only provide a brief description of these areas.

6.3 Spend Analysis

This term is a general umbrella term used to capture various strategic activities that are important for designing a sourcing strategy for the corporation. The steps involved in developing a sourcing strategy are:

1. **Data Warehouse for Spend Analysis**: This requires the creation of a homogeneous data warehouse from disparate (heterogeneous) databases (from various departments, locations). A formidable challenge to doing this is that the transaction records in different systems across the enterprise are not cross indexed. An additional challenge is that these records might refer to the same supplier by different names. This requires the use of data cleansing and data scrubbing tools. Therefore, some of the subtasks to creating a data warehouse are:

   a) *Supplier Normalization*: This entails the creation of a list of distinct suppliers so that transactions to the same supplier (e.g. IBM) but referred to differently in different systems (e.g. IBM NA,

I.B.M., etc.) can be mapped to a unique supplier, say International Business Machines. Moreover, the parent child relationships often need to be resolved – this is particularly difficult since M&As often lead to parent-child relations within companies that are completely different.

b) Commodity Mapping: This requires each transaction to be mapped to an appropriate commodity code, such as the UNSPSC code or a company proprietary code. This is due to the fact that the transaction records at the invoice level often provide only a part level description of the commodity and maybe associated supplier codes.

c) Data Visualization: Once the data is scrubbed and cleansed, a set of visualization and rendering tools are required to view the different cross sections of the data so as to get an enterprise-wide view of procurement spend.

2. **Sourcing Strategy**: Once a data warehouse is available for analyzing the procurement spend, the next step is to evaluate the different sourcing options for each commodity class or other dimension and identify the potential cost savings. This then provides a basis for a list of actionable sourcing initiatives. The subtasks to creating such a strategy report as follows:

a) *Demand Aggregation*: The data warehouse provides a means to examine spend by each category, supplier, plant etc. An important first step is to establish the number of suppliers being used for each commodity class across all plants. Often such an exercise might reveal that the number of suppliers that are being used for each commodity is very large and it presents an opportunity to allocate the demand to a few suppliers and leverage the aggregate demand volume to negotiate better prices.

b) *Supplier Scorecarding*: While consolidating the supplier set for any commodity class it is important to analyze the supplier performance against a set of company's strategic metrics. The score carding function helps identify the top suppliers to whom future allocation awards would likely go (despite potentially having higher prices) as well as the bottom suppliers who would need to be more aggressively managed as part of the "supplier relationship" activities.

c) *Catalog buying*: For each commodity class identify the total procurement cost of historical transactions against the catalog prices of a set of potential suppliers. This comparison provides a means to identify the cost savings of aggregating spend and using one or few suppliers with a set of broadcast catalog prices. A important difficulty in doing this is that the part numbers described in the transaction records need to be mapped to the corresponding part in a catalog that will not be (in general) indexed by the same commodity or part number codes.

d) *Contract Pricing*: Another aspect of a sourcing strategy is to decide on a negotiation technique for contract pricing. For example what are the potential cost savings of using auctions versus RFQs and how do these savings depend on the auction format. The ability to provide good estimates depends on how well the cost types of the suppliers can be characterized. In addition, it is important to model and analyze the risks associated with the uncertainty in the demand and choose contracted volumes optimally.

e) *Report Generation*: Finally a report needs to be generated that outlines the sourcing strategy based on the spend analysis.

6.3.1   Functional Requirements

There are three basic functional requirements for enabling the features that have been outlined above:

1. **Data Cleansing/Scrubbing**: An important ingredient for analyzing spend data is to get an integrated view of the heterogeneous transaction data that exists in various parts of the enterprise. This is central to supplier normalization, commodity mapping, and catalog buying. Often the same commodity or supplier is referenced differently and there does not exist a common key across these tables that allows for easy mapping. Currently the data cleansing activity is largely a human activity and the realm of the expert consultants. There are increasing efforts to provide tools that help to semi-automate this process based on two approaches:

   a) *Text Similarity:* The first technique is based on using text analysis (such as text distance metris). Such techniques provide automated mapping with an accuracy of about 50-60%.

   b) *Machine Learning:* In the presence of training data (where human experts provide mappings) machine learning techniques can be used to improve the performance to over 90% for the same class of data.

2. **Data Warehousing**: A data warehouse that provides an integrated view of all the relevant data from disparate sources is critical for generate various views of aggregated spend and report generation. Data warehousing is a well developed technology and most sourcing platforms provide some level of support for warehousing scrubbed data.

3. **Analytics**:   Developing a strategic sourcing report requires several types of analysis ranging from data mining to optimization. Few platforms provide the entire breadth of analytic capabilities required to generate strategic sourcing report and largely depend on catalog price based comparisons.

   a) *Data Mining/Statistical Analysis*: is necessary to determine supplier types from past transaction and/or bid data to establish the cost types of a supplier pool.
   b) *Optimization*: to determine the optimal contract levels to manage demand uncertainty to estimate cost savings from contract pricing.

Table 6.1 summarizes a mapping from the required features (rows of the table below) for spend analysis to the functional components (columns of the table) that are required to enable these features.

6.3.2   Data Cleansing/Scrubbing Techniques

One of the simplest and most common techniques for doing cleansing tasks, such as supplier name normalization, involves the use of text similarity methods. Such methods have long been used in various fields such as information retrieval and molecular biology, and a rich repository of literature exists describing a large number of such methods and variations thereof (Salton and Buckley, 1987; Navarro, 2001; Baeza-Yates and Ribeiro-Neto, 1999).

Table 6.1. Mapping of Features (Rows) to Functional Components (Columns)

	Data Cleansing	Warehousing	Analytics
Supplier Normalization	X		X
Commodity Mapping	X		X
Data Visualization		X	
Demand Aggregation	X	X	X
Supplier Scorecarding		X	X
Catalog Buying	X		X
Contract Pricing			X
Report Generation		X	

A frequently used technique measures the similarity between different strings by means of some sort of a distance function, whereby low distance values imply "more" similar strings (with zero distance implying complete similarity) and high distance values imply correspondingly dissimilar strings. One such distance metric is the Levenshtein Distance (LD) (Levenshtein, 1966). Also referred to as edit distance, it is equal to the minimum number of character insertions, deletions and/or substitutions needed to convert one string into the other. Thus,

> LD("IBM", "IBN") = 1 since one substitution is needed to transform IBM to IBN
>
> LD("Delphi", "Delta") = 3, since two substitutions and one deletion is needed to transform Delphi to Delta.

LD is fairly robust to spelling errors and small local differences between the strings. Here, we give a dynamic programming formulation of the algorithm to compute LD between two strings, s1 and s2, where $|s1|=n$ and $|s2|=m$. In this formulation, computation of LD(s1, s2) requires the successive calculation of each cell of an array C[0..m, 0..n], starting from C[0,0], where C[i, j] denotes the distance between the first 'i' characters of s1 and first 'j' characters of s2. Each value, C[i,j] is calculated on the basis of the three cells, C[i,j-1], C[i-1,j] and C[i-1,j-1], along with the costs of the different operations (addition, deletion or substitution). Formally,

Initialization:

$$C[0, 0] = 0$$
$$C[i, 0] = C[i-1, 0] + 1, \; 1 \leq i \leq m$$
$$C[0, j] = C[0, j-1] + 1, \; 1 \leq j \leq n$$

Calculation: Compute C[i,j] for all $i \geq 1, j > 1$ using the formula

$$C[i, j] = Min(C[i-1, j-1] + C_m, C[i-1, j] + 1, C[i, j-1],+ 1),$$

where $C_m = 1$ if $s1[i] \neq s2[j]$ and 0 otherwise.

While the above formulation assumes unit cost for substitution, and deletion/insertion operations, extensions have been proposed to handle different operation costs, transpositions etc (Navarro, 2001).

Another kind of string similarity method breaks the strings up into a set of tokens, and compute the distance between the strings based on these tokens. Strings can be tokenized in a variety of ways. One way is to use a word-based tokenizer that splits the string into tokens based on white space and punctuation. Another way is to use 'n'-grams, where n is a positive integer. In this case, each consecutive substring of 'n' characters is treated as a token. Thus,

- Using a word-based tokenizer, "The New York Times" would be broken into a set of 4 tokens, {"The", "New", "York", "Times"}.
- Using 'n'-grams with n=4, the "The New York Times" would be tokenized as {"The ", "he N", "e Ne", "New ", "ew Y",...}

Strings represented as such sets of tokens can be compared by applying various token based similarity measures. One such measure is simply to compute the number of terms in common between the two strings. The higher the number, the more similar the two strings are. However, this metric favors longer strings, and as such, normalized versions of this metric are often used. One such metric is the so called Jaccard distance (Jaccard, 1912), which is the number of terms in common divided by the total number of unique terms in the two strings. Thus,

Jaccard distance: $JD(s1, s2) = |S1 \cap S2| / (|S1 \cup S2|)$

A more general method is the tf-idf (Term Frequency – Inverse Document Frequency) (Salton and Buckley, 1987) approach where each token is assigned a weight representing the importance of that term within that particular string as well as relative to all other strings to which it is compared. Though this measure is more of a form of indexing textual strings or documents, it can also be used to measure similarity between different strings (or documents) in a given set of strings. In this case,

Term Frequency (tf) = # of times token occurs in the string
Document Frequency (df) = # of strings in which that token occurs
Inverse Document Frequency (idf) = $\log(N/df)$

Then, each token, t_i in the string is given a weight,

$w_i = tf_i * idf = tf * \log(N/df_i)$

where N is the total number of strings being compared. This way tokens that are frequent in a string but rare in the collection are given a higher weight. Similarity is then computed as a cosine distance (CD) between the vectors of the two documents

$CD(s1, s2) = \sum w_i \bullet w_j$

where w_i and w_j are the weights of the ith and j^{th} terms of s1 and s2. Several different variations of this have been studied as well (Salton and Buckley, 1987).

While such similarity measures provide a useful way of doing such cleansing tasks as supplier name normalization and commodity mapping, they can be computationally expensive since distances may have to be computed between all pairs of strings in a relatively big universe of strings (e.g. the set of all suppliers). This is more of a problem when edit distances are used since dynamic programming in inherently very expensive. Thus, in practical situations, it may be more worthwhile to use such measures in conjunction with clustering and/or rule-based approaches. One such approach, discussed by McCallum, Nigam and Ungar (2001), divides all strings into weak overlapping subsets, called canopies, using some fast but cheap measures. Example of such measures could be the Jaccard distance (or its unnormalized form) discussed above, or the tf-idf metric in conjunction with the cosine distance for only parts of the strings (say first token only). Once the canopies are formed, then the more expensive methods such as LD can be used to compare the strings within each canopy to determine sets of similar strings (clusters).

While the string similarity methods described above are sufficient for certain kinds of cleansing activities, such as supplier name normalization, they are often not enough for other tasks, such as commodity mapping. In such cases, machine learning techniques are often helpful in improving the quality of the results. Since data cleansing for spend analysis involves mapping and manipulation of textual data, fields such as information retrieval and natural language processing offer a plethora of machine learning techniques that have been found effective in such domains (e.g. maximum entropy (Nigam, Lafferty, and McCallum, 1999), support vector machines (Joachims, 1998) and Bayesian methods (McCallum and Nigam, 1998)). Detailed discussions of these approaches are out of scope of this article and the interested user is referred to the above mentioned references. Nevertheless, we do discuss in the following section how these techniques can be used in the data cleansing activities needed for spend analysis.

6.3.3 Data Cleansing for Spend Analysis

As explained previously, two main cleansing activities often have to be carried out before an aggregate view of the procurement spend across the organization can be developed using the transaction data, namely supplier name normalization and commodity mapping. While the techniques described in Section 6.3.2 can be used to perform these tasks, they alone are often not enough to do these well and must be augmented with other approaches. Below, we describe how these cleansing techniques can be used specifically for supplier name normalization and commodity mapping.

Supplier Name Normalization

The approach followed for correctly mapping different names for the same supplier (from disparate data systems/transactions, etc) to a common unique name depends upon the availability of a unique (normalized) set of suppliers for the enterprise. If such a list is already available, then the task boils down to comparing the supplier associated with each transaction to the set of suppliers in the normalized list and identifying the closest match.

Moreover, transactional records often contain demographic data such as address and contact information for suppliers that can, and should, be used to aid the normalization exercise. While string similarity measure described previously can be used on the supplier names alone to do the normalization exercise, the quality of the results will be greatly enhanced by using such additional information. While data like zip codes and phone numbers can be directly matched, the string similarity measures can be used to match the addresses associated with the various suppliers. By combining such similarity matches on supplier names along with exact and similarity matches on demographic data, it is possible to get a higher degree of success in the normalization than by using the supplier names alone. This exercise can be further enhanced by making use of simple techniques such as stemming, stop word elimination as well as other common filtering approaches (removing special characters, transformation of numbers to uniform format, etc). Matching of addresses, however, can introduce some complications in the process due to different formats used across various systems, especially if the address is available as a string rather than in attribute-value form. In such cases, extracting various attributes such as zip codes, street addresses and city may require the use of manually created regular expressions or rules, or in some cases, automatically created rules using machine learning techniques like RAPIER (Califf, 1998) or Winnow (Zhang, Damerau and Johnson, 2002).

If a normalized list of suppliers does not exist, then a clustering exercise is needed in conjunction with the above mentioned approach to break down the list of all the suppliers that exist in the transactions into sets of names where the names in each set belong to the same supplier. To do the clustering, once can use the canopy method outlined in the previous section. To create canopies, cheap methods such as zip code matches, phone number matches and name and/or address similarity matches using tf-idf can be used. Once the canopies have been formed, the more expensive techniques such as the Levenshtein distance can be used for performing similarity matches across supplier names and street addresses.

Commodity Mapping

Commodity mapping requires each transaction to be mapped to an appropriate category from either a standard classification code (such as the UNSPSC code) or a company specific code. Transactional data can have commodity information from several different sources such as part de-

scriptions, general ledger descriptions, invoice descriptions, SIC descriptions, purchase order descriptions and accounts payable descriptions. Any
or all of these descriptions may be misspelled, incomplete or absent. On
the other hand, the commodities, to which the transactions need to be
mapped to, are often arranged in a taxonomy, and are often accompanied
by textual descriptions of each such commodity. The task is then to map
the textual data from the various description fields in the transactional data
to the name and/or textual descriptions of the commodities in the target
taxonomy. Currently, much of this aggregation is done manually where
millions of transactions are mapped to corresponding commodities by the
use of filtering rules and manual matches of various transactional descriptions to the descriptions associated with the commodities in the target taxonomy. Not only is this method cumbersome and error prone, it is also de-
void of any fixed methodology that can be consistently repeated for future
mapping exercises since it is highly subjective and non-algorithmic. However, the use of machine learning techniques can automate this process
with a high degree of accuracy and also provide a consistent, algorithmic
framework for doing such mappings repeatedly with newer transactional
data and/or commodity hierarchies.

In this section, we discuss some of the techniques available for automating the task of mapping transactions to related commodities using machine
learning and string similarity methods. The actual approach taken depends
on the availability of historical transactional data that has already been
mapped to the corresponding commodities.

If such historical data is available, then one can avail of a large number
of machine learning techniques that are commonly used for classification
tasks involving textual data. As described previously, several different approaches can be considered, including maximum entropy (Nigam,
Lafferty, and McCallum, 1999), support vector machines (Joachims, 1998)
and Bayesian methods (McCallum and Nigam, 1998). Irrespective of the
actual approach adopted, two steps are involved: (i) learning classification
models for predicting commodities based on textual information from
transactional data, and (ii) applying these models to the unmapped transactions to determine the appropriate commodities for those transactions.
Several issues arise during these two steps.

First, the textual fields from the transactions have to be tokenized and
converted to instances for learning to take place. For tokenization, various
methods, as discussed in the previous section, can be used. Similarly, several methods can be used to convert the sets of tokens into instances for
learning. One common approach is to consider the set of all tokens (in the
entire dataset) as a set of features, and convert the textual data on each
transaction into an instance of binary-valued features where a feature takes
a value of 1 if the corresponding token exists in the token set for that

transaction, and 0 otherwise. This, however, results in a fairly large feature set and correspondingly sparse data. With the extremely large number of transactions that often need to be mapped in an enterprise this often leads to a highly computationally expensive learning process. Moreover, the high dimensionality of the resultant dataset often degrades classification performance due to overfitting. As such, it is often necessary to use feature selection to eliminate redundant and/or useless features. Several different approaches can be used for this purpose. A simple approach would be to use some of the token weighting schemes discussed in the previous section such as token frequency, inverse document frequency or their product. Then, tokens whose weights are above (or below) a certain threshold (depending upon the metric used) are discarded and the remaining subset is used for learning the models. For example, one could remove all tokens whose term-frequency is below a certain threshold under the assumption that very infrequent tokens are not useful for classification. Other approaches use statistical tests, such as chi-squares, to determine which features are more relevant. Yang and Pedersen (1997) provide a comparative study of these and other feature selection methods.

Second, a decision has to be made regarding the level of the commodity taxonomy for which the classification models must be learned. The ultimate target, the 'commodity' level of the taxonomy, also corresponds to the highest level of difficulty with respect to the classification task, since a classifier has to select one out an extremely large target set (the set of all commodities). On the other hand, if models are built for classification at a higher level of the taxonomy, the number of target classes is significantly reduced, thereby improving the performance of the classifiers as well. For example, the UNSPSC code has several levels such as segment, family, class and commodity. As such, following a hierarchical modeling and mapping approach will generally be better than attempting to predict commodities directly. Thus, instead of learning classifiers for directly predicting at a lower level (e.g. commodities), it is generally better to build classifiers for first predicting at a higher level (e.g. segments). Additional classifiers are then built for predicting lower levels, given that the higher level value is already known. As such, in the case of the UNSPSC code, classifiers would be built to predict the product family given its segment, product class given its segment, and eventually associated commodity given the product class. Thus, classification would be carried out in a hierarchical manner with the highest (most general) class first (e.g. segment), followed by lower (more specific) classes (e.g. family, then class, then commodity). One does not always have to start at the highest level, nor stop at the lowest level. It depends to the type of data at hand, the number of potential classes (commodities) to which the data has to be mapped, and the specificity needed for the task at hand (i.e. whether spend needs to be computed at the commodity level, or at the class level, etc.).

If, however, historical mapped data is not available, then one needs to create such mapped data for training the classifier models. Once such data has been prepared, the method described above (for mapped data) can be used to do the commodity mapping exercise. Consequently, the available transactional data is split into two parts. One part (smaller, say 10%) is then mapped to the appropriate commodities, and then used to train the classification models. The remaining data is then mapped to the appropriate commodities using these classification models. In order to create the mapped data needed for training, one case use the clustering and string similarity methods described in the previous section. As in the case of supplier name normalization, fast, cheaper methods such as tf-idf and simple rules can be used to loosely group together similar transactions into canopies, and then use more stringent measures such as edit distances can be used to refine the clusters further. Each such cluster is mapped to an appropriate target (segment or commodity etc. depending upon the level for which models have to be learned), once again using string similarity methods. Irrespective of the methods used, this step will invariably require human intervention to make sure the mapping is being done correctly, otherwise the errors in the mapped data will propagate to the rest of the transactions as well during the learning and classification phase. One can alleviate this issue by using human experts only in cases where more than one target class has a high likelihood of being the correct map.

6.4 Sourcing

6.4.1 Core Ingredients

A core ingredient of sourcing is *negotiation* with suppliers and executions of the resulting contract. There are three critical components required for negotiations:

1. **RFx**: The RFI (Request for Information), RFQ (Request for Quotation), and RFP (Request for Proposal) – collectively referred to as RFx – each represents a document and a means for a buyer to specify the requirements of a purchase along multiple dimensions from multiple suppliers.
2. **Protocol**: for negotiations that define the process by which the negotiation is conducted.
3. **Contracts**: The negotiations lead to a contract which is then executed with one or more suppliers.

6.4.1.1. RFx

In B2B settings, the specification of purchases can get quite complex and require sophisticated capabilities that allow the specification of complex items or services.

Complex RFQs also need to allow for a variety of bid structures that exploit complementarities and economies of scale in cost structures of suppliers.

An RFx is a document with an associated process initiated by a buyer in order to solicit information, competitive quotes, or proposals from multiple suppliers. A sourcing platform should make the RFx process as easy and straightforward as possible for all of the parties involved. It should also be versatile enough to be used for both goods and services, and for both direct and indirect spend categories. It should also support a wide range of RFx types and sizes, from simple RFIs to complex RFPs.

Most RFx applications support a common set of capabilities such as the creation and editing of an RFx document that mimics its paper-based counterpart. For example, this includes being able to add any number of questions with response fields of the attribute types expected by the buying organization (e.g., numeric, date, text, units of measurement, etc.) for each line item.

Table 6.2 summarizes some of the major requirements in terms of bids (from the seller side) that are supported.

After receiving such bids the buyer needs to identify the set of bids that minimizes total procurement cost subject to business rules such as:

- The number of winning suppliers should be greater than a certain number (to avoid depending too heavily on just a few suppliers), but smaller than a certain number (to avoid too much administrative overhead);
- The maximum amount purchased from each supplier is bounded to a certain limit;
- At least one supplier(s) from a target group (e.g., minority) needs to be chosen; and
- If there are multiple winning bid sets, then one needs to pick the set that arrived first.

Decision support capabilities are essential to facilitate the creation and evaluation of such complex RFQs and bids. Identifying the cost-minimizing bid set subject to these business rules is a hard optimization problem and difficult to do by hand (as is a common practice today). In addition, the buyer is required to specify a scoring function that specifies the tradeoff of the non-price attributes against price. This is difficult to do in a consistent manner without a rational process to elicit the tradeoffs.

6.4.1.2. Protocols (aka Auctions):

A typical flow for negotiation is to get a bid response to the RFQ from the suppliers and choose the appropriate bid/s that satisfy the requirements of the purchase at minimum cost. With the advent of the Internet, online reverse auctions represent a new tool in the purchasing department's toolbox to potentially increase competition through open, real-time, competitive bidding, which requires an iterative bidding protocol.

Table 6.2. Description of Complex Bid Types

Bid Types	Description
Simple multi-line bids	A bid includes multiple items, and specifies the unit price for each item.
Multi-attribute bids	A bid includes multiple items, and specifies various relevant attributes for each item, including unit price.
Bundled bids	A bid includes multiple items, specifies the quantity of each item, and provides a total bid price for all the items.
Volume discount bids	A bid includes multiple items, and specifies the price curve of each item.
Configurable bids	A bid includes multiple items, and specifies various relevant sets of values for each attribute for each item. This provides a compact representation for a large number of configurations (e.g. PCs) and needs to support mark-up based pricing.

RFQs are often used in a single round process that is similar to a one shot sealed bid auction where the winners are selected (based on the recommendations of the bid evaluation engine) once all the bids are in. However, in a price negotiation context, it is often desirable to have a multi-round process where after each round the suppliers are allowed to reformulate their bids based on information about the winning bids (more like based on feedback from the auctioneer). Such a multi-round process is illustrated in Figure 6.2. The bid evaluation engine provides the decision support for all the three functions required for multi-round negotiations and iterative auctions. Winner determination identifies the winning bids from a given set of bids to minimize the total procurement cost, the pricing module prescribes the payment to be made by each winner (this could be in general different from the bid price to promote efficiency in the market), and signaling provides a "market clearing" price for bid reformulation. This iterative process continues until there are no new bids or closing time.

A reverse auction is similar in many ways to an RFx. For instance, it is a process initiated by a buyer in order to solicit competitive bids from multiple suppliers. In fact, some vendors view RFx's and reverse auctions synonymously. However, there are important differences with the most prominent being that most auction formats allow for live, real-time, open, competitive bidding where bidders must outbid the current winning bid in order to win the business. Of course, there are a variety of auction formats and settings. The most basic reverse auction is a price-only auction for a single item. Most auction providers (and there are many) provide a wide range of formats and settings including multiple quantities of a line item, multiple line items, time extensions, start and reserve prices, partial quantity bids and award allocations, and bundled bids – to name only a few. However, there are three advanced auction formats/settings of note:

- Combinatorial Auction – allows suppliers to mix bundled bids along with un-bundled bids
- Volume Discount Auction – allows suppliers to establish price discounts at certain quantities
- Multi-attribute Reverse Auction – the winner(s) is determined by a score (rather than just price) calculated using the buyer's weights and preferences for price, quantity, and any number of other attributes (similar to an RFx but using open bidding)

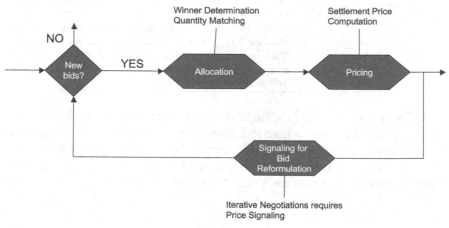

Fig. 6.2 Process Flow for Iterative Auctions

6.4.1.3. Contract Management:

One of the main goals of a sourcing project is to execute one or more purchasing contracts with one or more suppliers. The prices and terms of the line items covered by a contract were previously negotiated in RFx and auction rounds. The contracts themselves, however, also go through a different form of negotiation at a more legally precise level. Once executed, these contracts are meant to be used to procure the contracted line items (perhaps via a procurement system) using the negotiated prices and terms. The general idea, of course, is to have as much spend as possible purchased under contract (under the assumption that these prices and terms are better than what's currently available in other markets). A sourcing platform should provide a means to generate a contract based on its preceding RFx and auction negotiations, support contract negotiations, and monitor compliance to the contracts' business commitments over time.

A number of contract monitoring capabilities are required. The simplest of these is an alert notification sent when a contract is soon to expire. More advanced contract monitoring capabilities help ensure vendor compliance as well as buyer compliance. For example, the buyer's purchase volume commitments can be monitored with alert notifications sent if there is danger of buying under the minimum quantity within the designated time period. Likewise, notifications can be sent alerting the buyer and/or supplier of a supplier's violation of a delivery commitment.

In all of the contract management solutions today, there is a specific and important shortcoming. Namely, there are two key parts enabling automated contract monitoring which currently must be performed manually. First, the contract commitments to be monitored must be manually culled out of the contract's negotiated legalese into a structure amenable to analysis. This is analogous to the upfront structuring of RFx's into explicit attribute types so that they may be more easily evaluated and scored. Ideally, the contract negotiations themselves would use a more structured mechanism to negotiate and record business commitments to enable their automated monitoring. The second manual part enabling contract monitoring is the capturing of the business process data and raw transaction data needed to assess whether commitments are being fulfilled or violated. Ideally, this data is captured automatically through probe points and other IT system instrumentation, and integrated with the contract management system.

The ideal contract monitoring scenario described above represents nontrivial enhancements to a contract management system – the first part due mostly to cultural and behavioral challenges, the second part due mostly to technical challenges. Despite their challenges, these enhancements are needed if monitoring contractual commitments that impact business performance is to be taken seriously and if there are many contracts to monitor. Business process integration and management (BPIM) and business activity monitoring (BAM) systems are beginning to address these challenges.

6.4.1.4. *Decision Support for Bid Evaluation:*

A suite of decision support tools, are required for bid evaluation:

- Tools to elicit a buyer's preferences for multi-attribute bid evaluation based on conjoint analysis and advanced decision analysis techniques;
- A visualization tool to compare multiple bids across different attributes; and
- A bid evaluation engine that uses optimization techniques to identify a cost minimizing bid set subject to various business rules.

Multi-Attribute Bids

Most procurement negotiations include non-price attributes over and above price. A typical RFQ for office chairs is shown in Figure 6.3.

Buyers need to take a number of different factors into account when evaluating and selecting bids. For example, there may be factors related to the *product specification* such as price, material quality and properties, color and size. In addition, there may be factors related to the *service specification* such as delivery time and cost, and warranty. Furthermore, there may be *supplier qualification* factors such as trading history, experience and reputation.

Decision support systems are needed to evaluate and score weighted preferences of multiple attributes. Besides traditional scoring mechanisms, tools that allow users to interact with the system to determine weights of multiple attributes from ordinal rankings of subsets of submitted bids are required. The bidders should also be allowed to describe their multi-attribute bids as a set of attribute values, but they can also specify complex pricing rules for product configurations. Also, it provides an advanced interactive visual analysis capability that allows buyers to view, explore, navigate, search, compare and classify submitted bids.

		Buyer Specifications	Bids from Suppliers		
Office Chairs	Price	Reservation Price ($50/unit)	$54/unit	$36/unit	$45/unit
	Quantity	500	500	350	500
	Delivery	7-10 days	7 days	10 days	8 days
	Reliability	> 70% (Probability On-time)	100%	Unknown	80%
	Quality	Med-High (Ergonomics)	Hi	Med	Med

Fig. 6.3. RFQ for a Multi-Attribute Negotiation

A utility function can be used to design a (strategic) scoring function that communicates to the suppliers how the buyer will evaluate multiattribute bids. One approach to eliciting the preference structure is to model it as a polyhedra in the space of the parameters and design pairwise comparisons to quickly narrow down the feasible region. In order to perform this in real time, we need efficient techniques for estimating the centroid and a cut that is perpendicular to the "longest axis" of the polyhedra thereby minimizing the feasible region for the parameters. In this paper, we present the use of a "hit-and-run" algorithm for sampling uniformly from a polyhedra. We tailor the use of this algorithm to also produce a cut that approximately minimizes the volume of feasible polyhedra. The advantage of this technique is its relative simplicity - it relies only on matrix algebra and avoids the use of nonlinear optimization techniques. Computational results suggest that this method is fast and accurate (Ghosh and Kalagnanam 2002).

Preference aggregation concerns how to combine the preference rankings of a number of suppliers by multiple buyers during strategic sourcing and the combining of search results from multiple. For large contracts and governments contracts it is often necessary the ranking process be transparent to a larger audience. Some interesting approaches for generating a consensus ranking based on ideas of social choice theory have been studied by Davenport and Kalagnanam (2004).

Bundled "All-or-Nothing" Bids

While negotiating prices for procuring, say, weekly demand, it is advantageous to aggregate demand over several locations and plants, because this leads to a larger transaction. An additional advantage is that suppliers can provide a discounted bid on a bundle (e.g., a demand for sugar in New York and in New Jersey) because they might have excess inventory in a local warehouse or spare capacity in the carrier and hence can reduce transportation costs. However, the discounted bid price is valid, only if the entire bundle bid is accepted.

In such settings, finding the cost-minimizing bid set ensuring that the demand for each item is satisfied can be a very hard problem as the number of bids begins to get large. (Notice that each supplier is usually allowed more than one bid and as the number of items increases the number of bids can get quite large. Also, notice that the optimal supply may oversatisfy demand.)

Volume Discount Bids

Volume discount bids allow the seller to specify the price they charge for an item as a function of quantity that is being purchased. For instance, a computer manufacturer may charge $1000 per computer for up to 100 computers, but for more than 100 computers would charge $750 per computer. Bids take the form of *supply curves*, specifying the price that is to be charged per unit of item when the quantity of items being purchased lies within a particular quantity interval.

Procurement Item - Sugar	Seller1	Seller2	Seller3	Seller4
100 in NY	30	80	100	30
5 in CT	0	5	5	0
20 in NJ	20	10	20	10
Price submitted by Seller	$150	$125	$300	$125

Decision Variable   x1   x2   x3   x4

```
Minimize   150 x1 + 125 x2 + 300 x3 + 125 x4
Such that   30 x1 +  80 x2 + 100 x3 +  30 x4 >= 100
                      5 x2 +   5 x3              >=   5
            20 x1 +  10 x2 +  20 x3 +  10 x4 >=  20
```

Solution: x2=1 and x4=1 → Price = $250

Fig. 6.4. Example of Bundled Bids and Resulting Optimization Problem

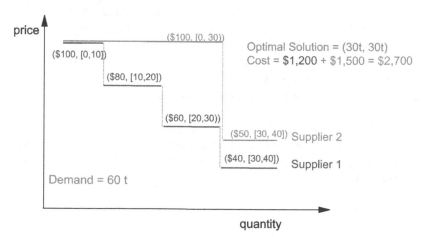

Fig. 6.5. Volume Discount Bids

When there are multiple suppliers providing such volume discount bids, the choice of the winning bids and the amount to be procured from each supplier is a difficult optimization problem that is modeled as an integer program. In addition, the various business rules are captured as side constraints within the mathematical formulation. Davenport and Kalagnanam (2001) provide a detailed discussion of the mathematical formulations for bid evaluations.

6.4.2 Support Functions

Some support functions that are useful for negotiations and contracts are (i) document management, (ii) project management, (iii) market intelligence, and (iv) collaboration and workflow.

Document Management

Files of all types – terms, drawings, schematics, etc. – are important documents for strategic sourcing. Additionally, the sourcing documents themselves – RFx's, auctions, projects, contracts, proposals, etc. – need to be well managed. These documents tend to have lifecycles, access controls, versions, sub-sections, sub-documents, and are often linked to one another and other attachments. A strategic sourcing platform must have an easy and scalable means to manage file attachments and sourcing documents.

For example, an RFx document has a lifecycle (e.g., creation, open, closed, deletion and archiving amongst others), has collaboration needs such as access controls for collaborators and simultaneous editing support, and several parts including header information, a schedule, a list of commodity code areas, a list of line items and questions organized by sections, file attachments, document references, and threaded discussions. Many of these parts including its lifecycle, schedules, file attachments, document references, and threaded discussions are managed by a common document management framework and available to all document types. For example, the RFx process typically spawns a set of Proposals submitted by suppliers which are themselves documents that get directly linked to the RFx document and managed along with it. Each Proposal has a lifecycle and can have file attachments, threaded discussions, etc.

Project Management

Most strategic sourcing activities are managed as projects with certain team members, goals, milestones, tasks, and schedules. And like most other sourcing activities, project management is a collaborative activity. A sourcing platform should provide some form of project management as a means to coordinate and schedule all of a sourcing project's activities and monitor the project's progress according to milestones and goals. Also like an RFx, a project can have threaded discussions and any number of file attachments. Projects can also include the list of other sourcing documents that comprise the project such as all of the RFx's, auctions, and contracts.

At the heart of project management are roles, tasks, and a schedule – a means to create milestones and tasks with owners and durations, each of which may be optionally dependent upon another task or milestone (e.g., "start this task when that task ends"). Schedule updates need to propagated automatically while the base schedule is still available for direct comparison. Furthermore, a project may include any number of sub-projects whose schedules can "bubble up" to the master project's schedule providing a complete schedule for the master project. This also goes for RFx's and auctions; their schedules can "bubble up" to the master project schedule. In fact, the main project's tasks and milestones can be dependent upon a task or milestone within any sub-project, RFx, or auction schedule. Ideally, projects can be made into templates for reuse.

Market Intelligence

A variety of external information helps category managers and other purchasing staff members perform their jobs. This external information includes commodity prices, supplier news, company news, and market intelligence such as industry news and reports.

Market intelligence needs to be provided as a sourcing workbench (a.k.a. a dashboard or enterprise portal) as the first page upon logging into the system. Like "My Yahoo!", this workbench can be customized in format and content to meet the information needs of each user. Portal functionality is useful for building such dashboards.

Collaboration and Workflow

Strategic sourcing involves many people. In addition to negotiating and communicating with multiple suppliers, category managers must collaborate with many people across the enterprise such as Engineering for technical specifications, Manufacturing for scheduling and inventory concerns, Legal for contractual terms and conditions, etc. A strategic sourcing platform must provide several means for all of these people and groups to collaborate effectively.

Just as with real documents, many people often contribute to their content, review, and approval. For example, an RFx can have an owner and any number of collaborators, each possibly responsible for a different part of the RFx with respective access permissions. A technical collaborator, for example, may be able to modify technical specifications for line items and evaluate each subsequent proposal based on its technical merits, but she may not be able to modify any terms and conditions which may be managed by a collaborator from Legal.

Another form of collaboration is threaded discussions. Each document needs to support threaded public and private discussions such as public discussions with suppliers about line item details, or private discussions with collaborators regarding requirements, evaluations, schedules, etc. All of these discussions should be maintained within the context of the RFx document for later review and archiving. Instant messaging is a useful complement to threaded discussions, and there are a number of applications today (most free) that are available.

6.4.3   Mapping Features to Functional Components

Table 6.3 below provides a mapping from the required features (rows of the table below) for spend analysis to the functional components (columns of the table) that are required to enable these features.

Table 6.3. Feature (rows) to Function (columns) Mapping for Sourcing

	Bid Evaluation	Preference Elicitation	Document Management	Collaboration & Workflow	Resource Allocation	Portal Server
RFx (Complex Bids)	X	X	X	X		
Auctions (Complex Bids)	X	X	X	X		
Contract Management			X			
Project Management			X	X	X	
Market Intelligence						X

6.5   Tactical Procurement

Procurement is a tactical activity where purchasing is performed within the umbrella of existing contracts. Typical purchasing within an enterprise starts with a requisition that is approved and purchased from within catalogs of selected suppliers. An additional activity that is supported at this level is the enablement of (new) supplier catalogs and the management of these catalogs.

Requisition systems provide support for placing an order for goods or services. They need to allow buyers to specify the items they wish to purchase, the quantities and the contracts that will govern the price and terms and conditions of purchase. In order to enable this, requisitioning systems need to support several features:

1. Workflow and Approval: Requisitions may need to be approved depending on business rules which are based on issues such as the amount of purchase, the types of items being purchased and the authority level of the buyer. For requisitions that are large in terms of dollar amount it may require multi-level approval from multiple parties. It also provides checking of inventory for item availability before a requisition is submitted and also ability of splitting requisition across multiple suppliers.

2. Contract & Inventory Management: Requisitions are governed by contracts and the purchases are accounted for against the contract obligations. In addition, the buyer should be able to issue a request for information about price and availability to suppliers before submitting the actual requisition.

3. Integration: Requisitioning requires information exchange with order and inventory subsystems, order processing and management. It also needs to integrate pricing, taxation, payment and fulfillment. An additional requirement is smooth interaction with backend systems such as ERP systems.

Catalog Systems

Catalog systems need to allow items to be organized in multiple category hierarchies. The items can be described with a rich set of extendable attributes and supports a number of browsing, product comparison and multiple search systems. Catalogs need to facilitate mass imported using XML formats and interface with of content management systems and to industry catalog and catalog management systems such as A2i.

Supplier Enablement

Supplier enablement is an important part of a procurement platform and should allow modeling of suppliers as multi-level hierarchical organizations. Individual users should be assigned a number of roles that, along with customizable access control policies, govern what individual employees of suppliers can do. Suppliers should be allowed respond to RFQs and requests for information. They can be allowed to own customized catalogs managed by the same catalog system.

6.6 Conclusions

This chapter provided an overview of the various functionalities that are required for procurement and sourcing for an enterprise. The goal was to outline these requirements and then provide a brief description of the technologies and the mathematics that underlie these technologies.

References

Baeza-Yates, R. and Ribeiro-Neto, B. (1999) *Modern Information Retrieval,* Addison Wesley

Califf, M.E. (1998) Relational learning techniques for natural language information extraction, PhD Thesis, University of Texas at Austin

Davenport, A.J. and Kalagnanam, J. (2004) A Computational Study of the Kemeny Rule for Preference Aggregation, *Proc. 19th National Conference on Artificial Intelligence* (AAAI-04)

Davenport, A. and Kalagnanam, J. (2001) Price Negotiation of Direct Inputs, in: Dietrich, B., Vohra, R. (eds.), *Mathematics of the Internet: Auctions and Markets*, Volume 127, Springer Verlag, New York, NY

Ghosh, S., Kalagnanam, J. (2003) Polyhedral sampling for multiattribute preference elicitation, *Proc. 4th ACM Conf. Electronic Commerce,* 256–257

Jaccard, P. (1912) The distribution of flora in the alpine zone, *New Phytologist* 11, 37–50

Joachims, T. (1998) Text categorization with support vector machines: learning with many relevant features, *Machine Learning: ECM-98, Tenth European Conference on Machine Learning,* 137–142

Levenshtein, V.I. (1966) Binary codes capable of correcting deletions, insertions and reversals, *Soviet Physics Doklady* 10 (8): 707–710

McCallum, A. and Nigam, K. (1998) A comparison of event models for Naive Bayes text classification, *AAAI-98 Workshop on Learning for Text Categorization*

McCallum, A., Nigam, K., and Ungar, L.H. (2000) Efficient clustering of high-dimensional data sets with application to reference matching, *Proc. of the sixth ACC SIGKDD International conference on Knowledge discovery and data mining,* 169–178

Navarro, G. (2001) A guided tour to approximate string matching, *ACM Computing Surveys,* 33 (1), 31–88

Nigam, K., Lafferty, J., McCallum, A. (1999) Using maximum entropy for text classification, *IJCAI-99 Workshop on Machine Learning for Information Filtering,* 61–67

Salton, G., Buckley, C. (1987) Term weighting approaches in automatic text re-
    trieval, Technical Report No. 87-881, Department of Computer Sci-
    ence, Cornell University, Ithaca, NY
UNSPSC, The United Nations Standard Products and Services Code,
    http://www.unspsc.org
Yang, Y., Pedersen, J.O. (1997) A comparative study on feature selection in text
    categorization, *Proc. of the 14th International Conference on Machine
    Learning ICML97*, 412–420
Zhang, T., Damerau, F., Johnson, D. (2002) Text chunking based on a generaliza-
    tion of Winnow, *Journal of Machine Learning Research*, 615–627

7 Managing Risk with Structured Supply Agreements

Colin Kessinger and Heiko Pieper

7.1 Introduction

Sourcing teams commonly manage spend equal to 30-70% of their firm's revenue. In addition, they function as the boundary between their firm and its supply base, requiring them to build and execute sourcing strategies in an environment where demand, supplier performance, pricing, and material availability constantly change. The magnitude of the dollars at stake in sourcing decisions can lead even small percentage miscalculations in either price or quantity to have dramatic effects on a company's margins, top line performance (through lost sales) and balance sheet (through inventory). Double-digit percentage reductions in stock prices, nine-digit misses in revenues, and ten-digit inventory write-offs are all well documented and recurring events attributable to mismatches between supply and demand. Given the magnitude of the uncertainties present in sourcing decisions, as reflected in the typical forecast errors in material requirements and supply conditions, avoiding such miscalculations is both extremely challenging and extremely valuable.

When it comes to managing risk and flexibility in your supply chain, it is all about reducing the time it takes to position assets, such as capacity or inventory, and then maximizing the revenue earned on those assets. Of course, in the absence of considerable supply and demand uncertainty, the time pressure on the supply chain would reduce considerably and the risk of having invested too much or too little would all but disappear. Unfortunately most business tools and approaches take a limited view of the uncertainty problem; for example relying only the point forecast as a measure of the demand, and rules of thumb to offset the effects of uncertainty. From leader to laggard we have found that this rule-of-thumb approach significantly underperforms in most business scenarios and even amplify the effects of forecast error. More bluntly, point forecasts and rules of thumb have cost companies billions in terms of lost market share, expedite fees, and inventory carrying costs, write-downs, and write-offs.

As a result, efforts are well underway to develop a more rigorous and comprehensive framework for quantifying and managing the effects of uncertainty. The Supply Risk and Flexibility Management (SRFM) framework focuses on risk-adjusted Total Sourcing Cost metrics, quantifying the performance of supply agreements (contracts) against a range forecast. The range forecast captures not only the low and high scenarios, but also the dynamic nature in which it might oscillate between the high and low scenarios. The approach also emphasizes risk-metrics, not just a static average or the "if-everything-goes-according-to-plan" projection. For example, most VMI/SMI programs projected zero inventories for the buyer on average or by plan, but resulted in considerable inventory liabilities when the forecast melted. A primary purpose in creating forward-looking risk metrics is to identify and then mitigate exactly these types of exposures.

In the remainder of this chapter we will review current processes and practices for building flexibility and explore the processes and tools required to identify supply exposures, to define flexibility requirements, and to implement risk sharing arrangements with suppliers, and assess tools and skills required to deploy Supply Risk & Flexibility Management business process.

7.2 Current Practices

There are four primary efforts in place to try and improve flexibility and reduce sourcing related risks. These are 1) internal efforts to align information and incentives from marketing and sales through order fulfillment, 2) sharing of information with the supply chain through visibility tools, 3) flex contracts, and 4) leadtime reduction initiatives.

In many organizations the silo problem persists; limited information is "thrown over the walls" from one function to the other and the organizations often have conflicting incentives. In the supply chain realm this problem begins with marketing having a wealth of information, but being forced to whittle it down to a single point forecast. In many organizations, there may be pressures to hit a certain number to meet financial goals. Sales, of course, wants this number as large as possible to ensure that operations can deliver enough material. Operations, however, has no visibility to all of the marketing scenarios and has no idea what changes were made to "make the numbers", and has no idea by how much sales inflated their projections. But they do not want to be caught short, so they often inflate the numbers again. Today, some companies are focusing on process redesign to minimize some of these issues; however, they also continue to rely on the point forecast as focal point, and, thus will see limited results from their efforts because the forecast is as much a business decision as it is a projection of demand, and because it still omits the information regarding the range of outcomes that are likely.

Considerable investments have been made into visibility technologies with the hope that they will help increase flexibility and minimize risks. One form of visibility is the sharing of forecasts. This really is an extension of the internal efforts described above; an effort to get the not only the company, but the entire supply chain aligned around one plan. Another form of visibility is that of the suppliers' production schedules and inventory positions. This real-time information enables buyers to understand their available-to-promise and identify gaps against customer orders as arrive. Many of these solutions have also started to support Sarbanes-Oxley in the US. These solutions certainly add considerable value by streamlining tactical execution and by enabling S-OX compliance, but fall short in the planning window of two to six months – the window in which most supply chain blow-ups occur. There are two underlying problems. First capacity levels are hard to compute and so dependent on mix that any snapshot of availability is likely to give a very misleading picture. Second, visibility does not mean control and control is what ultimately drives flexibility and risk management. For these reasons, it is not at all surprising that more formal contracts, in lieu of shared forecasts and visibility, emerge as a necessity during periods of constrained supply.

Flex contracts have been around in the form of flex-fences for at least a decade if not longer. These typically are defined as percentages up and down that volumes can change relative to the forecast. The percentages increase further out in time to reflect both the increased uncertainty and the supplier's ability to react in that amount of time. While seemingly a simple, one-size-fits-all, evergreen/auto-pilot solution, the devils for flex fences were in the details. First, the contracts assume that all parts require the same level of flexibility. In practice different parts and products simply have very different flexibility requirements. Second flex-fences assume upside and downside requirements remain constant through the product life-cycle and the business cycle. Nothing could be further from the truth; certainly your business objectives change from launch to EOL. Common sense even says there is more downside when demand is at historical highs than when they are at historical lows. And finally, third, the actual implementation, while simple on the surface, requires a lot of ifs, ands, and buts. As a result they often appear in contracts but rarely appear in practice. In the academic research, authors have developed complex analytical models to address some of these concerns, but the nature of the solution immediately eliminates the "seemingly simple, one-size-fits-all, evergreen/auto-pilot" attraction that led to the implementation of the program.

Finally, there are the lead-time reductions initiatives. Without a doubt these are worthwhile endeavors that yield significant gains. The primary challenge here is trading office an easily measured and tracked metric of price versus the hard to quantify value of a lead-time reduction. It is exactly this type of question that SRFM strives to answer.

While all of these initiatives strike some of the root causes, there are several key fundamental weaknesses along the lines of information, incentives, and control. To address these issues, the necessary information (business objectives and sourcing uncertainty) needs to be captured and terms of trade to meet the business goals by quantifying their impact on a forward looking basis need to be developed. The inability to quantify the effects of uncertainty and to develop supplier relationships that best match the "uncertainty landscape" with the firm's business objectives fundamentally limits the supply chain's ability to reduce the "total costs" associated with material prices, shortages, and inventory. The severity of this limit is such that at least 5% of the "total costs" are left on the table as unclaimed opportunities. Capturing this 5% will be the next evolution of state-of-the-art procurement.

7.3 Current Research

The problem that we solve is best characterized as finite-horizon, multi-mode, and capacity constrained with serially correlated stochastic demand and prices. While the intent of the broader framework is to drive superior sourcing performance, the model and algorithm value any "portfolio" of contract options by identifying and executing the optimal exercise policy. While a subset of this model and its outputs are very similar to the theoretical and computational research of base-stock policies, the contribution is best understood in the context of the capacity planning and supply contracts research.

The generic capacity planning problem is one of trading off economies of scale versus the risk of either building too much capacity or at least the cost of building too early. The basic problem consists of determining the future expansion times (or investment times), sizes, and locations, as well as type of production facilities in a way that optimize a particular strategy such as increasing market share, maximizing profit, or minimizing cost, etc.). In specific instances of the literature, demand may either be deterministic or stochastic, the cost of expansion is a concave continuous function, and the time to expand may be instantaneous or involve a finite lead-time.

Since 1950 cost minimization methodologies have been developed mainly on the field of operation research. Good surveys of this evolution appear in Manne (1967), Friendenfields (1981), Luss (1982), and Angelus(1997) as well as Leiberman (1989). Where the literature assumes deterministic demand, the cost models and expansion opportunities tend to have much more complex structures. For example they may include dependencies within the scale and sequence of expansions. Erlenkotter (1969, 1976) and Giglio (1973) are examples of these models.

Capacity expansion models with the objective of profit maximization have been one of the main streams of research around real options. A good review of this literature is provided by Dixit and Pindyck(1994). Here the typical model assumes a Markovian or semi-Markovian stochastic process and no lead time with a cost function that depends on the cumulative investment. Giglio (1970) and Erlenkotter (1977) extend these basic models by assuming more general cost functions and stochastic lifetime for the capacity. Finally Angelus and Wood (1996) consider the same cost structure as Giglio, but they introduce a geometric Brownian motion model for demand and optimize assuming access to only one capacity expansion.

Turning out attention to the contracts literature, there have been several efforts to extend the inventory literature by examining the impact of different contractual terms that typically accompany the lead-time constraint imposed by the supplier. For example, Bassok and Anupindi (1997) model the retailer's decision in a multi-period finite horizon framework in which the retailer receives a price discount for having made a minimum commitment over a finite horizon. In our framework, we will refer to this as a "swing" contract. Bassok and Anupindi (1995) and Tsay and Lovejoy (1999a) examine a rolling horizon problem in which the retailer can revise, within a contractually negotiated percentage, the commitments for each period within the horizon. This type of contract is the flex-fence policy in the current practice section. In both the work on the swing contract and the flex-fence contract, the authors solve the problem heuristically.

Another extension to the inventory literature examined situations with more than one supply source where the sources offered both different prices and lead-times. Zhang (1996) considers the problem with two and three supply modes under a periodic review policy. She also considers a situation where the more flexible supplier has a capacity constraint, perhaps reflecting a supplier's limited capacity to expedite. Moinzadeh and Nahmias (1998) consider a similar two-supply-mode problem, but under a continuous review policy. Finally Fisher and Raman (1996) examine a two-period model in which each period represents a portion of the selling-season and the buyer uses the second, shorter lead-time supply source to capitalize on the information learned from the initial sales of the product.

Finally there is considerable literature on contracts and their ability to achieve channel coordination. Although there are a few exceptions, the literature almost exclusively addresses stylized two period models that focus on managerial insight rather operational models. For those interested, Cachon (2004) provides a very recent review of the supply chain coordination literature.

Across the board, the literature has demonstrated the substantial opportunity of developing richer supply relationships. This, while not even tapping all of the potential. Most notably, most of the literature assumes that both parties are risk-neutral. However in practice there are significant opportunities to allocating the physical and financial risks to the party that is best able to manage it. Consider for example the swing contract explored by Bassok and Anupindi. Using a contract to guarantee that a small supplier can hit minimum revenue targets may help the supplier attain better financing terms on the assets needed to ensure delivery. In return, the supplier may give the buyer very aggressive pricing on any upside. In this case, the contract is creating value by eliminating disastrous outcomes in some scenarios in exchange for better terms in other scenarios.

7.4 Framework for Analyzing Structured Supply Agreements

In the following, we describe the framework of Structured Supply Agreements. The framework is defined as a set of five capabilities. Together with the computational risk analysis methodology described later, this enables the identification and implementation of supply agreements that significantly reduce the cost and risk for the supply chain.

The five capabilities are:

1. The ability to collect, communicate and respond to new information as soon as it becomes available.
   - Firms with this capability enable both themselves and their supply base to leverage the new information as it becomes available.
2. The ability to assess the impact of uncertainty about material requirements and supply performance on future sourcing performance.
   - This visibility takes the form of scenario-based analysis of the future sourcing and financial performance that result from a supply strategy.
   - Firms with this capability use it to set performance objectives, guide planning decisions, and manage expectations and performance risk
3. The ability to design supply strategies to achieve sourcing objectives across uncertain business outcomes
   - Firms with this capability use it to manage the impact of business uncertainties on their sourcing and overall business performance,

and to provide more specific guidance on their requirements to their suppliers.

4. The ability for management to efficiently and consistently shape performance commitments from suppliers and liabilities to suppliers, in accordance with their business objectives.
   - Firms with this capability use it to manage their strategic supply base as an extension of firm so that the "supply service" meets the top and bottom line objectives of the income statement, the asset and liability objectives of the balance sheet, and the customer service and business risk objectives of the firm's overall strategy.

5. The ability to monitor your supply position to proactively identify and address shortage or excess inventory exposures.
   - Firms with this capability use it to ensure that the flexibility that they put in place is executed to meet their business objectives and that they have sufficient early warning to address looming gaps instead of having to react to actual gaps.

This framework strives to achieve two fundamental objectives, 1) ensuring that your supply base can meet your objectives, and 2) ensuring that your supply base will meet your objectives. That the supply base can is a function of the quality of information that is shared. That the supply base will is a function of the control over the assets that you have in your supply chain. Both are a function of having the business and financial objectives aligned across the supply chain. Once a company can capture the uncertainty it faces and can develop metrics to measure and manage this uncertainty, structured contracts provide the necessary information, control, and alignment to drive the investment in capacity and inventory to meet the upside while balancing the buyer's and supplier's cost structure and risk profile in the flat or down markets.

7.5 Technical Implementation

7.5.1 Range Forecasting

The first step in calculating cost and risk of sourcing alternatives is the capturing of uncertainty as a basis for the analysis. In our framework, this is accomplished by the *Range Forecast*. A Range forecast captures the demand and price uncertainty in the future based on market conditions. This error can be reflected through forecast scenarios (high, base, low) scaled to the level of error commonly experienced in forecasts. However, this is more than the traditional what-if analysis. No demand pattern follows the smooth lines typically found in what-if analysis. Rather true demand is likely to oscillate in between and around the scenarios. Capturing this volatility is also essential as it significantly can change the performance on a sourcing contract. This volatility can be captured through the analysis of historical demand patterns and forward-looking scenarios. The stochastic model underlying the Range Forecast then translates these inputs into scenarios that preserve serial correlation, accurately reflecting and updating changes in the conditional distributions. The range and volatility of an example are shown in the following samples for an analysis of a new suspension module in the automobile industry.

While in this particular instance, a mean reverting/diverting stochastic process is used, the model and solution methodology is independent of the underlying stochastic process. Once the uncertainty about the key determinants of sourcing performance has been captured, the future-sourcing performance, by each business scenario of any given sourcing strategy can be calculated.

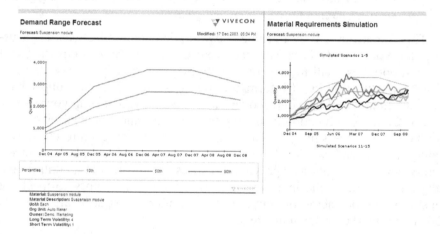

Fig. 7.1. Range Forecast Output

7.5.2 Computational Risk Analysis of Portfolios and Options of Structured Agreements

In the following, we will describe the framework of Structured Supply Agreements. It is in many ways an extension of the research described earlier. The main difference is that it extends the set of metrics for analysis, enabling deeper analysis of tradeoffs between the key elements of performance: price, availability, and liabilities/inventory. Together with the mathematical model to evaluate these agreements under uncertainty presented below, it enables the identification and implementation of optimal supply agreements that minimize cost and risk for the supply chain.

Structured Agreements capture and communicate constraints and performance tradeoffs throughout the supply chain. Each structured agreement can include one or more of the following components:

1. Quantity: minimum and maximum quantity in units for each period
   The buyer commits to ordering at least the minimum quantity in each period. In return, the supplier commits to the availability of the maximum quantity and the replenishment and capacity requirements connected with this commitment. Both sides often solidify the commitments through incentive payments or penalties if commitments cannot be met

2. Price: Unit price by period, fixed or market dependent
   This price can be changing over time based on a price schedule, it can be linked to a market index, and have potential caps and floors to limit market exposure to both parties.

3. Lead time: Order placed in period t will be delivered in period t + LT
   This is the leadtime (LT) for delivery by the supplier. A unit ordered in period t will be delivered in period t+LT. The reduction of leadtime generally enables the reduction of inventory and shortage risk, but leads to higher cost and liability exposure.

4. Cancellations of orders
   Orders that have been placed in period t can be cancelled in periods before t+LT. The extra cost and liability by this added flexibility is covered by a time-dependent cancellation fee paid by the buyer.

5. Swing Contract
   Minimum and maximum quantities are set over a range of periods. These are often in addition to period by period constraints.

6. Minimum order quantities (batch size)
   Many industries have minimum batch sizes for production that drive minimum per-order quantities. In many cases, these are significant cost and risk drivers in the supply chain.

7. Sequential capacity expansion options
   Sequential capacity expansions can be handled in many different ways. In our framework, capacity expansions are introduced as a new agreement that has a lead time and an associated payment for the capacity expansion cost.

The state space for uncertainty will be defined by the triplet (demand, price, availability). This information is captured in a single vector ω. Additionally in each period there is a separate event that determines the supplier specific availability dependent of demand and market conditions. We assume that this information is available at the time the order is placed.

The sequence of events is as follows:

1. Based on the uncertainty realized at the end of the prior period, inventory and inventory position is updated.
2. Uncertainty for the period is realized. This includes the realization of environmental uncertainty in addition to supplier specific uncertainty that drives supplier specific availability issues.
3. New orders are placed
4. Cost for the period is calculated.

7.5.3 Mathematical Formulation

Order Decisions

X_t matrix that contains all outstanding order information for period t. The rows (indexed by j) represent the different leadtime options (referred to hereafter as contracts) and columns (indexed by i=0 to LT) represent the number of periods until delivery. Contract 0 represents inventory.

$x^t_{j,i}$ quantity of units ordered on contract j due to arrive period t+i. This follows from the above definition

$x^t_{j,LT}$ decision variable for each contract 1 to J in period t

E_t matrix of indices that contains the ongoing expansions for period t. The rows (indexed by j) represent the different expansion options and columns (indexed by i=0 toT) represent periods when the expansion becomes available.

$e^t_{j,i}$ index (0,1) of expansions j available in period t+i. This follows from the above definition

$e^t_{j,LT}$ decision variable for each expansion 1 to J in period t

Z_t matrix of cancellations for contracts $j = 1 \ldots J$ in period t

$z^t_{j,i}$ (decision variable) is the quantity of units cancelled on contract j due to arrive period t+i.

Pricing Information

$p^t_{j,k}$ price for one unit at price tier k for contract j in period t.

$T_{j,k}^t$ quantity break points for the tiered pricing structure.

Availability

$\Delta_j^t(\omega_t)$ fraction of units that the supplier failed to deliver on contract j.

This is both a function of the state that describes the overall uncertainty as well as a "coin-toss" that determines the supplier specific availability.

Uncertainty

ω_t is the state of uncertainty (demand, price, availability). The state of uncertainty refers to the state of information at the time the decision is made.

Cost Parameters

S_t shortage cost in period t

BO_t back-order cost in period t

H_t physical holding cost in period t

bo fraction of shortages that are backordered.

Other Contract Parameters

$inv_{j,t}$ required investment to add expansion j in period t

$bp_{j,i}$ buyer penalty for canceling an order on contract j, i periods in the future

sp_j supplier penalty for not delivering an order on the schedules delivery date

$Max_j^t(\omega_t)$ maximum order quantity on contract j in period t. This parameter can be a function of the market conditions.

$Min_j^t(\omega_t)$ minimum order quantity without penalty on contract j in period t.

$Min\_batch_j^t(\omega_t)$ minimum order batch size on contract j in period t.

Max_sw_j maximum order quantity on swing contract j

$Max\_sw_j^t$ maximum order quantity on swing contract j in period t. The cumulative quantities for contract j are counted against the cumulative available units.

Min_sw_j minimum order quantity without penalty on swing contract j

$\beta_{j,i}^t$ fraction of order that can be cancelled on contract j when the order is i periods out

7.6   Calculation Engine

$$f_t(X_t,\omega_t)=\min\left[\begin{array}{l}S_tU(X_t,\omega_t)+BO_tV(X_t,\omega_t)+H_tO(X_t,\omega_t)+PC(X_t)+CAP(E_t)\\-\sum_j SP_j(X_t,\omega_t)+\sum_j BP_j(Z_t,x_{j,LT}^t)+\alpha E(f_{t+1}(X_{t+1},E_{t+1},\omega_{t+1}))\end{array}\right]$$

where

$$y_j = x_{j,0}^t\Delta_j^t(\omega_t)$$

-   Actual number of units received based on units expected to arrive this period minus units the supplier was unable to deliver

$$U(X_t,\omega_t)=\left[D_t(\omega_t)-\sum_j y_j\right]^+$$

-   Number of units short in period t

$$O(X_t,\omega_t)=\left[\sum_j y_j - D_t(\omega_t)\right]^+$$

-   Units of inventory after demand is realized.

$$V(X_t,\omega_t)=bo\left[D_t(\omega_t)-\sum_j y_j\right]^+$$

-   Number of units back ordered in period t

$$PC(X_t)=\sum_j\sum_k(p_{j,k}^t(\omega_t)-p_{j,k-1}^t(\omega_t))(y_j-T_{j,k-1}^t)^+$$

- Total purchase costs in the period

$$CAP(E_t) = \sum_j inv_j e^t_{j,LT}$$

- Investment cost for expansions in period t

$$SP_j(X_t, \omega_t) = sp_j x^t_{j,o}\left(1 - \Delta^t_j(\omega_t)\right)$$

- Supplier penalty for units not delivered at the time of delivery

$$BP_j(Z^t_j, x^t_{j,LT}) = \sum_j \left(bp_{j,LT}\left(x^t_{j,LT} - Min^t_j\right)^+ + \sum_{i=1..LT-1} z^t_{j,i} bp_{j,i} \right. $$
$$\left. + \left[bp\_sw_j\left(\sum_t x^t_{j,LT} - Min\_sw_j\right)^+ \right] \right]_{\text{If } t=T}$$

- This term is the buyer penalty based on failure to meet minimum purchase requirements plus cancellation penalties for units already ordered. Additional penalties are possible for not meeting minimum quantities for swing contracts.

Subject to:

$$0 \le z^t_{j,i} \le \beta_{j,i} X^t_{j,i}$$

- Constraint on the number of units that can be cancelled

$$x^{t+1}_{j,i} = x^t_{j,i+1} - z^t_{j,i+1} \quad \text{for } i = 0 \text{ to } LT\text{-}1, j = 2 \text{ to } J$$

- Updating of outstanding orders

$$x^{t+1}_{0,0} = \left[\sum_j y_j - D_t(\omega_t)\right]^+ - bo\left[D_t(\omega_t) - \sum_j y_j\right]^+$$

- Updating of inventory (contract 0) for next period

$$x^{t+1}_{0,0} \le Max^{t+1}_0(\omega_t)$$

- Maximum constraint on inventory given the state of uncertainty

$$x^t_{j,LT} \le e^t_{j,i} Max^t_j(\omega_t)$$

- Maximum order size on contract j in period t given the state of uncertainty and expansion

$$x^t_{j,LT} \leq Max\_sw^t_j(\omega_t)$$

- Maximum available quantity on swing contract j in period t given the state of uncertainty

$$x^t_{j,LT} \geq \delta^t_j Min\_batch^t_j(\omega_t), \delta^t_j \in (0,1)$$

- Minimum order batch size on contract j in period t given the state of uncertainty

The sourcing strategy that optimizes expected Total Sourcing Cost is given by the minimum of the objective function $f_t(X_t,\omega_t)$ subject to the constraints outlined in the model. The mathematical problem is an instance of a stochastic dynamic program with uncertainty given by the vector ω_t. The objective function is piecewise linear with mixed integer constraints due to the presence of capacity expansion and/or minimum order constraints.

The computational challenge of solving this model is given by the potentially large state space due to uncertainty in demand, price and availability combined with a large number of decision and state variables in the model (orders placed, backorders, inventory, and expansions). Integer constraints obviously further add to the difficulty. To appreciate the computational complexity, consider the problem of projecting the sourcing performance over the next 12 months in monthly increments. Assume the sourcing portfolio includes 2 sourcing alternatives and 1 capacity expansion option with a 3 month lead time. Formulated as a dynamic programming problem with 12 stages, each stage involves solving for the optimal order and expansion decisions for multiple instances of the discrete state space. Although each of these problems involves up to 50,000 constraints and an equivalent number of variables, it can be solved efficiently with standard optimization software, especially in the absence of integer constraints. In contrast, with a traditional dynamic programming formulation finding the optimal decision for each state requires solving hundreds and potentially thousands of decision problems for each stage. Any naïve approach will render this approach intractable in any reasonable time frame.

Nevertheless, a carefully designed algorithm that balances the trade-offs between the dimensionality of the state and uncertainty space has been designed. An efficient implementation of this algorithm solves regular problems within a few minutes and takes just slightly longer for planning problems over longer horizons. The algorithm sits at the heart of an enterprise solution that delivers cost and risk analysis of sourcing alternatives for direct materials in different industries. The system comprises of efficient data input, the range forecast analytics described above and a large set of configurable reports that can be used within the system or exported.

7.7 Contract Design & Management Reporting

SRFM introduces an extension to the traditional landed cost model. The basic measures in the risk-adjusted Total Sourcing Cost calculation are "fully-loaded" price, inventory/liability costs, and shortage related costs. The fully-loaded cost should reflect most of the terms in the total landed cost model, such as freight and taxes, as well as volume discounts, price floors and caps, restocking fees, etc. This part seems pretty straight forward.

The key distinction in a risk-adjusted calculation is that fully loaded cost and additional metrics are evaluated over hundreds of forward-looking scenarios so that metrics such as the average inventory level, and average percentage short can be computed across a large number of scenarios. Furthermore true risk metrics such as the probability that inventory will exceed, for example, 90 days, or that shortages will exceed 10%, or that the backlog will exceed 1 month can also be computed. Price risks can also be projected. A buyer may want to evaluate the exposure to expedite fees on production at the suppliers or on freight or the exposure to price increases in a capacity constrained supply market. In companies where these metrics have been successfully introduced, management is specifying targets on both the average performance of the contract as well as performance against different risk metrics across a range of scenarios.

For example, to decide between 2 sourcing alternatives, expected Total Sourcing Cost over all possible scenarios is the key metric to focus on. Furthermore, the performance for specific demand scenarios, e.g, high or low demand might be taken into account, or a sourcing strategy with a slightly higher inventory risk in return for a reduced shortage risk might be preferred, even though Total Sourcing Cost is higher. Sometimes, more important than the average value, the timing of shortages or inventory risks is a key focus for the organization. In these situations, detailed over-time reports provide the tool to fine tune the strategy. While the automatic risk management system relies on numerical data, risk and its evolution over time is much easier visualized, so both numerical data and graphical reports are provided to support these functions. Further functionality enables the optimal utilization of sourcing alternatives in volatile business environments and acts as an early warning system. The system can be fully

automated, but also enables detailed sensitivity analysis to validate the results or solve non-standard problems.

In the following sections, we will showcase applications within different
industry sectors that utilize the described range forecasting and valuation
method.

7.8   Industry Examples

7.8.1   Tooling/Capacity Planning in Automobile Industry

In the last 3 years, several automotive manufacturers introduced a sunroof
integrated into an all glass roof. Clearly the adoption of this new roof was
highly uncertain; first it was more expensive than the traditional sunroof,
and second many customers may still prefer the traditional sunroof. In the
auto industry it is commonplace for the buyer to pay for the production
specific machines, tooling, fixtures, and gauges. Hence the buyer is faced
with making an investment that will determine the capacity level long before the adoption of the option is known. Therefore, the impact of the investment decision, in conjunction with the company's aggressive policies
regarding customer backlog, led to a critical trade-off between price risk
and availability risk.

Relying on a rule of thumb to cover the plan plus a standard percentage,
the buyers consistently found themselves over- or under-investing in capacity. In this case, the benefit of applying SRFM was threefold. First in
receiving a range forecast instead of a point forecast, the buyers knew what
range of outcomes they would need to cover. Second, knowing the range
of demand that they would likely need to cover, the buyer can evaluate a
range of strategies, factoring in the initial investment plus the cost and time
to expand capacity. Third, by quantifying the performance of these different strategies, the business objectives could be met at the lowest cost and
risk. Below is a sample of the output from the analysis (Figure 7.2). The
first alternative, 82k, corresponds to the rule of thumb for a high demand
scenario, and the abbreviation OT represents overtime and Exp represents
capacity expansion. Of course the capacity expansion required a considerable leadtime.

Without delving into all of the details, the take-away from the tables and
charts are that by decreasing the capacity investment and adding an option
for capacity expansion (62k+Exp), prices would have reduced by 6% and
9% in the low scenarios, 5% and 6% in the medium scenarios, and 4% and
0% in the high scenarios. Additionally (the colors correspond to percentiles) there is roughly a 15% chance that if the 62k option is selected the
capacity expansion will be required. Additionally, when the shortages oc-

cur, they almost always are less than 5%, with only a few periods seeing a small likelihood of reaching 10%. Given that consumers of this brand are willing to wait some amount of time for their vehicle, this backlog was consistent with retaining most of those customers.

Capacity Alternatives

Capacity	Investment	Low Demand		Med Demand		High Demand	
		Price	Shortage	Price	Shortage	Price	Shortage
82k	€ 120,000,000	€ 900	**0.0%**	€ 688	**0.0%**	€ 575	**4.3%**
70k+OT	€ 110,400,000	€ 851	**0.0%**	€ 656	**0.0%**	€ 557	**4.8%**
62k+Exp	€ 105,600,000	€ 828	**0.0%**	€ 647	**0.2%**	€ 576	**6.4%**

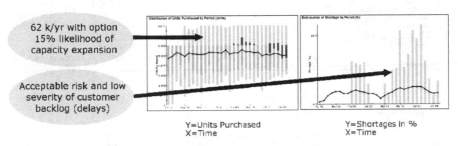

62 k/yr with option 15% likelihood of capacity expansion

Acceptable risk and low severity of customer backlog (delays)

Y=Units Purchased
X=Time

Y=Shortages in %
X=Time

Fig. 7.2. Tooling & Capacity Manager Report

7.8.2   Capacity and Launch Planning in the CPG Industry

Figure 7.3 shows some of these risk metrics in action for a CPG buyer. The report compares two launch plans and compares them over hundreds of demand scenarios. The report groups results into the lowest 25%, the middle 50%, and the highest 25% demand scenarios (note this grouping of hundreds of scenarios is distinctly different from running three scenarios). The top of the report shows a pro-forma cost statement for each scenario group, and the bottom provides explicit measures for service level and inventory performance.

There are several key insights from this report. First, some alternatives may be most attractive due to their performance in the low or high cases. For example, even if the new approach showed a 1% increase in Totals Sourcing Cost in the mid range (currently 2.4% better), the fact that it was 11.8% better in the low range (because it reduced the inventory write-offs from $346k to $234k) and 4% better in the high range (because it reduced shortages from 2.8% to 0.5%) may make the second approach preferable. Second, while performance along the mid-range may look pretty reasonable on inventory and shortage metrics, the exposures to inventory in the low case and shortages in the high case may be completely unacceptable.

Report Viewer - Microsoft Internet Explorer provided by Vivecon

Analysis Results   Sourcing Metrics by Scenario Group ▼

Sourcing Metrics by Material Requirements Scenario Groups

	A: Current Launch Plan (400M Capacity)			B: Flex LT=2mo $0.94/unit			Improvement ('	
Sourcing Costs ($NPV)	Low Range (25%)	Mid Range (50%)	High Range (25%)	Low Range (25%)	Mid Range (50%)	High Range (25%)	Low Range (25%)	Mid Range (50%)
Material Cost								
Material Purchases	778,206	875,830	1,691,422	664,594	851,426	1,895,129	14.6	2.8
Decrease in Inventory Value	-19,235	-33,059	-69,388	-10,028	-23,711	-65,214	-47.9	-28.3
Net Material Cost	758,972	842,771	1,622,034	654,565	827,715	1,829,915	13.8	1.8
Inventory Related Costs								
Financing Cost	18,644	23,493	17,780	10,028	13,343	12,376	46.2	43.2
Storage Cost	0	0	0	0	0	0	N/A	N/A
Total Inventory Costs	18,644	23,493	17,780	10,028	13,343	12,376	46.2	43.2
Shortage Related Costs								
Value of Material Short	N/A	N/A	N/A	N/A	N/A	N/A	N/A	N/A
Margin Lost Due to Shortages	N/A	N/A	N/A	N/A	N/A	N/A	N/A	N/A
Total Shortage Related Costs	0	2,291	358,728	0	2,807	69,890	N/A	-22.5
Buyer Penalty	0	0	0	0	0	0	N/A	N/A
Supplier Penalty	0	0	0	0	0	0	N/A	N/A
Other Payments (amortized)	180,000	180,000	180,000	180,000	180,000	180,000	0.0	0.0
Total Sourcing Cost	957,616	1,048,555	2,178,542	844,594	1,023,866	2,091,981	11.8	2.4
Price ($/unit)	Low Range (25%)	Mid Range (50%)	High Range (25%)	Low Range (25%)	Mid Range (50%)	High Range (25%)	Low Range (25%)	Mid Range (50%)
Average Material Purchase Price	0.90	0.90	0.90	0.94	0.94	0.94	-4.4	-4.4
Average Total Price	2.01	1.23	1.17	1.78	1.19	1.06	12.4	3.3
Quantities	Low Range (25%)	Mid Range (50%)	High Range (25%)	Low Range (25%)	Mid Range (50%)	High Range (25%)	Low Range (25%)	Mid Range (50%)
Total Units Required	525,543	940,587	2,188,998	525,543	940,587	2,188,998	0.0	0.0
Total Units Purchased	909,092	1,032,129	2,067,447	746,748	970,685	2,221,248	17.9	6.0
Availability of Supply	Low Range (25%)	Mid Range (50%)	High Range (25%)	Low Range (25%)	Mid Range (50%)	High Range (25%)	Low Range (25%)	Mid Range (50%)
Total Units Short	0	1,276	192,608	0	1,612	37,405	N/A	-26.3
Average Shortage (%)	0.0	0.1	2.8	0.0	0.2	0.5	N/A	-100.0
Inventory	Low Range (25%)	Mid Range (50%)	High Range (25%)	Low Range (25%)	Mid Range (50%)	High Range (25%)	Low Range (25%)	Mid Range (50%)
Average Inventory Value ($)	94,084	120,547	94,360	51,070	69,493	68,221	45.7	42.4
Average Inventory (units)	346,276	173,707	104,844	206,539	82,132	72,575	40.4	52.7
Average Inventory (days)	494	162	43	299	75	30	39.5	53.7
Value of Inventory Write-Off ($NPV)	347,606	71,594	0	234,287	21,022	0	32.6	70.6

Fig. 7.3. NPI Manager Report

Figure 7.4 shows even greater detail on the inventory story. The following graph shows the distribution of outcomes over all of the demand scenarios considered. The colors, as indicated by the legend, correspond to percentiles. The top of the gray represents the 90th percentile; 90% of the outcomes were below the top of the gray bar. The top of the dark blue bar corresponds to the 75th percentile; 75% of the outcomes were below the top of the gray bar. Revisiting the SMI example discussed throughout this article, this chart might only show a gray bar, suggesting that the 75th percentile of inventory was zero, but the top of the gray bar may show an exposure much greater than zero, just as in the last downturn when liabilities ballooned to 180 to 360 days of supply.

On the chart on the left hand side of Figure 7.4, we see a typical fashion goods launch strategy; positioning a large supply of FGI (in this case the black line shows that prior to demand a large buffer was installed) to fill the channel and capture the benefits of a successful product. In the remainder of the product life-cycle, the legacy of this risky positioning translates into substantial inventory levels. In contrast, the suggested alternative substantially reduces the inventory levels, on average and at each percentile.

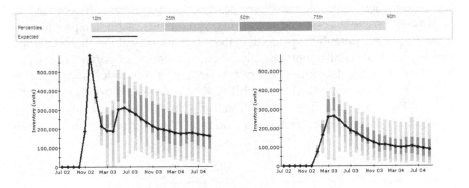

Fig. 7.4. Inventory Over Time Graph

7.8.3 Supply chain coordination in the High-Tech Industry

Post launch, the focus turns to managing A-Part spend. Typically these parts are high-value, long leadtime, exposed to allocated capacity, or available from limited sources. The primary purpose of SRFM in this context is balance the trade-offs between availability and liabilities while continuing to hit price targets. It goes without saying that long leadtimes are impediments to flexibility. Unfortunately business cycles can create problems out of even short leadtime components. Towards the end of 2000, many components were on allocation, prices were increasing, and shortages were rampant. In a matter of months, capacity utilization dropped to as low as 30% in some sectors, and months of inventory become quarters if not years of material. The write-offs and write-downs were well documented in the hundreds of millions. Today we already see this cycle repeating itself. For example, fabs (bare boards) went from rock-bottom prices to constrained supply in last six months of the year.

Consider the case of one electronics capital equipment manufacturer. The equipment is highly configurable, coming in over 10,000 possible configurations. Fortunately for most of their products, most of the supply challenge revolves around just 20 part numbers that account for nearly 70% of the cost on the BOM. Here the application of SRFM is twofold; first to negotiate flexibility terms commensurate with the uncertainly levels in this high volatile industry, second to monitor the ongoing supply position to proactively identify bottlenecks and to ensure balance across the commodities.

7.8.3.1. Create a Portfolio of Supply Sources

Figure 7.5 shows two sourcing "portfolios", each made up of a fixed quantity commitment at a price discount designed to serve baseline demand at the lowest possible cost, and some flex agreements at higher pricing designed to cover potential upside demand. Portfolio A attempts to balance the expensive flexibility option with the current option of purchasing material at the 3 month leadtime. Portfolio B drops the 3 month leadtime, relying entirely on a large upside capability that is considerably more expensive. While these are just two options, obviously there are many other alternative combinations. The key question is how much of each option to incorporate.

	Portfolio A	Portfolio B
	Quantity	Quantity
Firm Commit	800 5% discount	800 5% discount
Flex Option (3 mon LT)	2500	0
Peak Flex (1 wk LT)	1500 10% premium	4500 15% premium

Fig. 7.5. Contract Alternatives

These differences in contract structure generate important differences in each dimension of sourcing performance. Specifically, Portfolio A provides lower expected material cost, while the Portfolio B, with the large flex contract, reduces both inventory and shortage exposure. Which one is right for your business? Do you want superior performance in the lower scenarios or the higher scenarios?

7.8.3.2. Monitor Ongoing Supply Position

Below is an example of the supply position report used by the capital equipment manufacturer.

There are several noteworthy elements to this report. First, the report consists of forward-looking projections, providing a management level overview of the state of the supply over the upcoming months (the next year in this example – more likely the next quarter or two). The analyst can also provide month-by-month drill downs in case the average performance over the quarter or year does not reveal significant exposures in any particular month. Second, it reports both average performance as well as risk metrics as they are defined by the decision maker. As in the VMI/SMI example described in the introduction, the Y Channel satisfies the average inventory constraint (less than 40 days), but fails the risk-inventory constraint (less than 120 days). Third, the report highlights the material in violation of any of the management goals. Again this report is just exemplary. In the actual implementation additional metrics were reported, such as purchase level recommendations, cash outflow, and maximum supportable ship plans.

Report Viewer - Microsoft Internet Explorer provided by Vivecon

Analysis Results | Sourcing Metrics by Scenario Group ▾ | Add Report to Project... | Print...

Sourcing Metrics by Material Requirements Scenario Groups

Sourcing Costs ($NPV)	B: 800 Firm/2500 Flex/1500 Peak			C: 300 Firm/0 Flex/4500 Peak			Improvement (%)		
	Low Range (25%)	Mid Range (50%)	High Range (25%)	Low Range (25%)	Mid Range (50%)	High Range (25%)	Low Range (25%)	Mid Range (50%)	High Range (25%)
Material Cost									
Material Purchases	2,565,293	3,251,659	3,903,328	2,652,974	3,573,916	4,646,548	-3.4	-9.9	-19.0
Decrease in Inventory Value	-231,649	-112,951	-55,206	2,732	3,650	17,517	-101.2	-103.2	-131.7
Net Material Cost	2,333,644	3,138,908	3,848,122	2,655,706	3,577,567	4,664,065	-13.8	-14.0	-21.2
Inventory Related Costs									
Financing Cost	16,716	8,155	4,553	105	346	658	99.4	95.8	85.5
Storage Cost	238,617	115,314	63,206	856	3,377	5,337	99.6	97.1	91.6
Total Inventory Costs	255,334	123,469	67,760	960	3,722	5,995	99.6	97.0	91.2
Shortage Related Costs									
Value of Material Short	4,961	50,781	394,930	0	5,796	67,651	100.0	88.6	82.9
Margin Lost Due to Shortages	7,025	71,487	556,712	0	6,893	80,429	100.0	90.4	85.6
Total Shortage Related Costs	11,986	122,268	951,642	0	12,691	148,080	100.0	89.6	84.4
Buyer Penalty	0	0	0	0	0	0	N/A	N/A	N/A
Supplier Penalty	0	0	0	0	0	0	N/A	N/A	N/A
Other Payments (amortized)	0	0	0	0	0	0	N/A	N/A	N/A
Total Sourcing Cost	2,600,964	3,384,644	4,867,524	2,656,666	3,593,980	4,818,140	-2.1	-6.2	1.0

Price ($/unit)	Low Range (25%)	Mid Range (50%)	High Range (25%)	Low Range (25%)	Mid Range (50%)	High Range (25%)	Low Range (25%)	Mid Range (50%)	High Range (25%)
Average Material Purchase Price	169.80	172.58	174.54	192.83	194.22	195.14	-13.7	-12.5	-11.8
Average Total Price	190.18	166.52	220.13	192.89	195.09	201.12	-1.4	-4.6	8.6

Quantities	Low Range (25%)	Mid Range (50%)	High Range (25%)	Low Range (25%)	Mid Range (50%)	High Range (25%)	Low Range (25%)	Mid Range (50%)	High Range (25%)
Total Units Required	14,706	19,699	25,883	14,706	19,699	25,883	0.0	0.0	0.0
Total Units Purchased	16,105	20,094	23,859	14,691	19,648	25,421	8.8	2.2	-6.5

Availability of Supply	Low Range (25%)	Mid Range (50%)	High Range (25%)	Low Range (25%)	Mid Range (50%)	High Range (25%)	Low Range (25%)	Mid Range (50%)	High Range (76%)
Total Units Short	25	503	2,381	0	32	308	100.0	89.4	84.4
Average Shortage (%)	0.1	0.8	5.4	0.0	0.1	0.8	100.0	87.5	85.2

Inventory	Low Range (25%)	Mid Range (50%)	High Range (25%)	Low Range (25%)	Mid Range (50%)	High Range (25%)	Low Range (25%)	Mid Range (50%)	High Range (25%)
Average Inventory Value ($)	239,755	116,562	64,095	983	3,937	6,157	99.6	96.6	90.4
Average Inventory (units)	1,422	686	375	5	20	31	99.6	97.1	91.7
Average Inventory (days)	24	10	5	0	0	0	100.0	100.0	100.0
Value of Inventory Write-Off ($NPV)	0	0	0	0	0	0	N/A	N/A	N/A
Total Inventory Write-Off (units)	0	0	0	0	0	0	N/A	N/A	N/A
Ending Inventory Value ($)	261,220	127,336	63,590	0	5,046	1,168	100.0	96.0	98.2
Ending Inventory (units)	1,543	754	374	0	26	6	100.0	96.6	98.4

Fig. 7.6. Contract Alternatives Report

SUPPLY POSITION REPORT
Sourcing Performance Metrics

Reporting Period: 9/2003 - 8/2004	Projected Performance							
	Inventory				Write-Offs		Shortages	
	Expectation		Risk (10% chance value will exceed)		Average	Risk (10% chance value will exceed)	Risk 10% chance average value will exceed	
	$ OHB	DOS	$ OHB	DOS	$M	$M	units	%
Goal:		<40		<120				=0%
Interface	$623,700	12	$997,920	85	$16,444	$51,633	1	0.0%
Testhead A	$779,922	41	$1,247,875	99	$31,401	$98,598	1	0.0%
Testhead B	$834,036	22	$1,334,458	76	$12,494	$39,231	2	0.0%
Z Channels	$3,337,754	16	$5,340,406	67	$94,581	$296,983	1	0.0%
Y Channels	$5,527,120	23	$8,843,392	133	$365,000	$696,554	5	2.3%
Controller	$381,000	8	$609,600	55	$7,367	$23,132	1	0.0%
Power A	$494,700	13	$791,520	47	$1,453	$6,321	0	0.0%
Power B	$247,350	17	$395,760	75	$755	$2,370	0	0.0%
Clock board	$350,773	15	$561,237	77	$2,321	$7,288	2	0.0%
Cooler	$277,840	109	$444,544	195	$891	$2,798	2	0.0%
Memory board	$149,832	41	$239,731	121	$13,642	$42,836	0	0.0%

Fig. 7.7. Supply Position Report

7.9 Summary

History has shown that wherever assets meet uncertainty, the risk of multi-million dollar misses or even stock-price altering events is real. This happens at stages of the product life-cycle. The right processes, tools and frameworks for managing these risks can generate huge savings as well as protect the income statements and balance sheets from violent swings. Companies have long adapted sophisticated tools used by highly-trained professionals to manage currency risk. The time has come to apply the same resources to manage their sourcing risk in the increasingly outsourced environment.

In this chapter, we outlined the necessary steps to develop the processes, tools and framework. At the heart of these steps are the ability to capture the uncertainty that you are trying to manage and ability to project the performance of your initiatives against this uncertainty. This can be accomplished by the Range Forecasting techniques and the mathematical valuation model we introduced. As always the right set of metrics will ensure that you are asking and answering the right set of questions. The set of industry examples spanning the Automobile, CPG and High-Tech sector demonstrate the use and benefits of this approach.

While both the academic literature and the actual examples in this chapter demonstrate the considerable benefit of managing risk and flexibility, the adoption of these practices is not immediate. First and foremost, a sophisticated tool is required to quantify and manage the risk. However, to date only one vendor supplies such a tool and the costs to build a tool internally are substantial. Second, the application of this discipline is cross-functional. Not only are several functions required to participate in the process, but the metrics that are affected are spread across the income statement and balance sheet, and therefore, the entire organization. Therefore, considerable change management is required to align the different organizations and to introduce and track the new set of metrics.

References

Angelus, A. (1996) Optimal Sizing and Timing of Capacity Expansion with Implications for Modular Semiconductor Wafer Fabs, Working Paper, Stanford University

Angelus, A., Wood, S.C. (1996) The Effect of Capacity-Interdependent Costs on Size, Timing, and Value of Irreversible Investments, Unpublished Report, Graduate School of Business, Stanford University, CA

Bassok, Y., Anupindi, R. (1997) Analysis of Supply Contracts with Total Minimum Commitment, *IEEE Transactions* 29, 373–381

Cachon, G.P. (2004) Supply Chain Coordination with Contracts, Working Paper, The Wharton School of Business, University of Pennsylvania

Dixit, A.K., Pindyck, R.S. (1994) *Irreversible Investment with Uncertainty and Scale Economies*, Princeton University Press, Princeton, New Jersey

Erlenkotter, D. (1969) Preinvestment Planning for Capacity Expansion: A Multi-Location Model, Ph.D. Dissertation, Graduate School of Business, Stanford University

Erlenkotter, D. (1976) Coordinating Scale and Sequencing Decisions for Water Resources Projects, *Economic Modeling for Water Policy Evaluation, North Holland/TIMS Studies in the Management Sciences* 3: 97–112

Erlenkotter, D. (1977) Capacity Expansion with Imports and Inventories, *Management Science* 23, 694–702

Fisher, M., Raman, A. (1996) Reducing The Cost Of Demand Uncertainty Through Accurate Response to Early Sales, *Operations Research* 44, 87–99

Friedenfelds, J. (1981) Capacity Expansion: Analysis of Simple Models with Applications, Elsevier North Holland, NY

Giglio, R. J. (1970) Stochastic Capacity Models, *Management Science* 17, 174–184

Leiberman, M. B. (1989) Capacity Utilization: Theoretical Models and Empirical Tests, *European Journal of Operational Research* 40, 155–168

Luss, H. (1982) Operational Research and Capacity Expansion Problems: A Survey, *Operations Research* 30, 907–947

Manne, A.S. (1961) Capacity Expansion and Probabilistic Growth, *Econometrica* 29 (4): 632–649

Moinzadeh, K., Nahmias, S. (1988) A Continuous Review Model For an Inventory System with Two Supply Modes, *Management Science* 34 (6): 761–771

Tsay, A., Lovejoy, W. (1999) Quantity-flexibility Contracts and Supply Chain Performance, *Manufacturing & Service Operations Management* 1 (2): 89–111

Zhang, V. (1996) Ordering Policies for an Inventory System with Three Supply Modes, *Naval Research Logistics* 43: 691–708

8 Reverse Logistics – Capturing Value in the Extended Supply Chain

Moritz Fleischmann, Jo van Nunen, Ben Gräve and Rainer Gapp

8.1 Introduction

Conventional supply chain perspectives consider a set of processes, driven by customer demand, that convey goods from suppliers through manufacturers and distributors to the final customers. However, this is not where the story ends. Physical goods do not simply vanish once they have reached the customer. Nor does the value incorporated in them. Therefore, many goods move beyond the conventional supply chain horizon, thereby triggering additional business transactions: used products are sold on secondary markets; outdated products are upgraded to meet latest standards again; failed components are repaired to serve as spare parts; unsold stock is salvaged; reusable packaging is returned and refilled; used products are recycled into raw materials again.

The set of processes that accommodate these goods flows, which can often be interpreted as running 'upstream' in a conventional supply chain scheme, is known as 'reverse logistics'. Examples are manifold. Two categories, however, form the basis of the growing importance of reverse logistics throughout the past decade, namely return agreements for excess products and extended producer responsibilities.

The first category refers to a customer's right to return a purchased product and be refunded. Due to their increased channel power, retailers have been able to negotiate the right to return excess stock to manufacturers. Supply chain management analyses have shown that this type of return contracts can, in fact, be beneficial for both the manufacturer and the retailer. Thereby, the manufacturer's benefit hinges on larger expected sales volumes (Tsay et al. 1998). Similarly, consumers often have the legal right to return products within a certain period after purchase. This factor is gaining particular importance in the context of e-business, where customers cannot physically inspect products prior to purchasing. All of the above cases confront companies with returns of technically 'new' though possibly outdated products. Subsequent options differ by case. In the simplest case, products may simply be restocked. Other products may require repackaging or thorough inspection. Yet other products are salvaged through outlet channels. However, even in the case of simple restocking, effective administration and efficient handling of returns often constitute serious challenges.

The second category of reverse logistics activities that have drawn much attention is related to used products. Increasingly, companies are held responsible for the entire life-cycle of their products. By this token, several countries require companies to take back and recover their products after use by the customer. A well-publicized example concerns the recent directive on Waste of Electrical and Electronic Equipment (WEEE) of the European Union (see European Commission 2004). Even in less regulated environments, such as the U.S.A., increasing disposal costs drive companies to offering used product take-back as a customer service. At the same time, companies have been recognizing the value potential of used products. In particular, many high-end products from the business market are still valuable in other market segments, even after a few years' use. Similarly, used products may contain valuable components that can serve as spare parts. This value potential renders used products attractive not only for the original manufacturer but also for specialized third parties. In either case, this business requires novel supply chain processes that include the former 'user' as a 'supplier'.

In the past decade, reverse logistics has grown to a significant business sector. Most logistics service providers offer reverse logistics as one of their core competences. A quick search on the Internet yields links to a host of reverse logistics programs. Many leading original equipment manufacturers (OEMs) are engaged in product recovery initiatives and highlight them in their company reports.

At the same time, reverse logistics has also gained recognition in the academic community. Many leading supply chain management conferences feature dedicated sessions on this topic. The number of related articles published in academic journals has been growing exponentially. Several renowned international journals have recognized the topic through special issues (e.g., *Interfaces* 33(4), 2003; *Production and Operations Management* 10 (2+3), 2001; *California Management Review* 46(2), 2004). Recent books on reverse logistics include research monographs, textbooks, and case collections (e.g., Rogers and Tibben-Lembke 1999; Guide and Van Wassenhove 2003; Dekker et al. 2003; Flapper et al. 2004).

In this chapter, we review the field of reverse logistics. We discuss its opportunities and its challenges and indicate potential ways for companies to master them. We highlight what makes reverse logistics different from 'conventional' supply chain processes, but also point out many analogies, and we explain how both views can be integrated into an extended supply chain concept. As a basis for our discussion, we draw on two sources. First, we review key results from the academic literature. Second, we complement them with illustrations of reverse logistics practice at IBM.

Throughout our analysis, we take a supply chain management perspective and we emphasize the need for differentiation. The main lesson learned from supply chain management concerns the benefits of a holistic view: Rather than trying to optimize individual business processes separately, companies need to coordinate processes along the entire chain, based on their underlying common goal, namely satisfying customer demand. Applied to our field of analysis, this implies that decisions in reverse logistics should consider the entire scope ranging from the original customer, as the source of product returns, to the future market for these products. In the subsequent sections, we highlight how current business practice still deviates from this ideal, in particular by focusing predominantly on either the supply or the demand side. In fact, one can take the supply chain management impetus even one step further by considering the 'original' chain and the 'reverse logistics' chain together. In this view, reverse logistics simply becomes a particular set of processes in an extended overall supply chain. In the literature, this extended chain is often denoted as 'closed-loop supply chain' (Guide and Van Wassenhove 2003). In the subsequent sections we highlight implications of this extended view and its link with novel business models, such as the shift from a physical-product orientation to a service orientation.

A second aspect that we believe deserves particular emphasis is the distinction of different reverse logistics environments. It is not surprising that early reverse logistics literature focused on basic common elements, thereby leaning towards a 'one size fits all' approach. However, as the field is maturing, a more detailed view is in order. While sharing a common set of processes, different reverse logistics environments entail different priorities, different preferences, and different trade-offs and therefore require different managerial decisions. We highlight these distinctions in our discussion.

The remainder of this chapter is structured as follows. Following this introduction, Sections 8.2 and 8.3 consider supply and demand interfaces in reverse logistics, respectively. The next two sections address the supply chain design that links these interfaces. Section 8.4 focuses on location decisions, while Section 8.5 zooms in on temporal coordination of reverse logistics processes. Section 8.6 summarizes our view on this field. We start each section with a general discussion and then illustrate it with IBM practice.

8.2   The Supply Side – Reverse Product Flows

As a first step to highlighting opportunities in extended supply chains, we take a closer look at the supply side, that is, at the potential sources of 'reverse' product flows. In line with the discussion in the previous section, we interpret this notion rather broadly and consider all flows that surpass the conventional supply chain scheme, i.e. from suppliers via manufacturers and distributors to the customer. This view encompasses, in particular, the two cases highlighted in the introduction, namely returns of excess products to the previous supply chain member (also known as 'commercial returns') and returns of discarded used products (also known as 'end-of-life returns'). However, the scope of our discussion is much broader and also includes, e.g., reusable product carriers, such as pallets, and boxes, rotable spare parts, and leased equipment. In many cases, these products move to an upstream supply chain stage. Yet the terms 'returns' and 'reverse' are primarily symbolic and should not be interpreted as necessarily going back to the original sender.

In the literature, several schemes have been proposed for classifying this diverse collection of extended product flows. In a previous contribution, we have grouped these flows into five broad categories, namely (i) end-of-life returns, (ii) commercial returns, (iii) warranty returns, i.e. failed products submitted for repair, (iv) production scrap and by-products, and (v) reusable packaging material (Fleischmann 2001). De Brito and Dekker (2003) classify reverse flows based on two dimensions, namely the supply-chain stage at which they occur (production, distribution, or use) and the sender's reason for disposing of the product. They argue that returned products are either defective or their original purpose has become redundant. It is worth pointing out that the boundaries of the latter category are somewhat blurred since they actually refer to the owner's relative valuation of keeping the product versus disposing of it. If only the incentives are high enough, he will be willing to give up the product. For a given product, the challenge for companies engaged in reverse logistics is, of course, to identify those sources to which they have to offer relatively little incentives.

As for the sender, one can also distinguish different drivers for the receiving party. The literature commonly lists economic, commercial, and legal motives. The most obvious driver for acquiring products is their future market value. Alternatively, product take-back may be a customer-service element which supports sales in the original channel. At the same time, companies can often exploit such a policy to showcase themselves as environmentally conscious. Finally, companies may simply have the legal obligation to take their products back, as discussed in the introduction. It is worth emphasizing, however, that even in the latter cases companies may eventually find opportunities for exploiting returned products as valuable resources. We discuss this aspect in detail in the next section.

Another important insight from the literature concerns the fact that reverse product flows arise, in principal, at all supply chain stages (de Brito and Dekker 2003). Each stage in the process implies different product characteristics - and thereby a different market potential. Products returned at the final stage are, by definition, used. In contrast, products returned during the distribution phase are technically new – although they may be commercially outdated. Similarly, it is worthwhile distinguishing sources in the business market from those in the consumer market. Business markets typically offer larger volumes of homogenous, relatively high-end products, whereas products are much more dispersed in the consumer market. Obviously, these differences heavily impact potential costs and revenues. Again, we follow up on these implications in the next section.

To make the above theoretical concepts specific, we illustrate their implementation at IBM. The electronics industry has been one of the key sectors of reverse logistics developments. The combination of a huge market volume, short product life-cycles, and a potential of repair processes results in a large potential supply for reverse logistics. At the same time, this large volume also entails significant environmental concerns. Therefore, it is no coincidence that electronic products have been playing a prominent role in the discussions of extended producer responsibility, such as the WEEE directive (see above). The fact that life-cycles of electronic equipment are determined primarily by technological progress, rather than by physical failure, represents both an opportunity and a challenge. On the one hand, many 'end-of-life' products are still in good working condition and may therefore find another useful application. On the other hand, quick depreciation puts this option under significant time pressure (see Blackburn et al. 2004).

Recognizing the impact of product returns, IBM has bundled their management in a dedicated business unit Global Asset Recovery Services (GARS), a subdivision of its Corporate Finance organization. GARS worldwide operations are subdivided in three geographies, namely the Americas (North, Mid, and South), Europe, Middle East, and Africa (EMEA), and Asia Pacific. Together, these operations handled some 1.5 million units of returned equipment in 2003, of which the vast majority are personal computers (PCs), including non-IBM brands. High-end mainframes and mid-range server equipment account for a smaller fraction of the return volume. GARS operates primarily in the business market. Its core supply consists of end-of-lease equipment, which is owned by its mother organization Corporate Finance. Other fractions of returns stem from 'old-for-new' buy-back initiatives and from commercial returns by supply-chain partners. In a growing number of countries, IBM also offers its customers the option to return end-of-life used equipment. Returns from the consumer market play a more subordinate role and are primarily driven by legal requirements, such as the WEEE directive. In the next section, we discuss the potential value of these different streams.

In addition to returned equipment, IBM also manages reverse logistics processes for replaced components that can be repaired and used as service parts. For a detailed discussion of these processes we refer to Chapter 9 of this volume.

8.3   The Demand Side – Remarketing

Having considered the supply side of reverse logistics, we now turn to the other end of the chain, namely potential market demand. Arguably, this is the single most important factor determining the profitability of any reverse logistics program. Therefore, carefully and systematically considering potential options is vital. Creativity plays an important role at this point.

The most straightforward option is simply to resell the obtained products, possibly in a different than the original market segment. In this case, the reverse logistics chain essentially provides a broker function. While this alternative preserves a maximum of the original value added, its overall profitability may be limited. For products with a significant market value the original owner can be expected to claim at least a part of this value. This holds in particular for the consumer market where online market places, such as e-bay, facilitate extensive 'second-hand' trading. In the business market, the value added of the broker function as an intermediate between supply and demand tends to be larger. Furthermore, reselling may be a viable option for products that are still in the possession of the OEM, such as commercial returns and lease returns. In all of these cases, technological progress tends to put significant pressure on the throughput time.

Other market opportunities require additional value-adding steps, such as repair, upgrading to a more recent technological level, or even extensive remanufacturing. In a few cases, these recovered products are indistinguishable from the original products and therefore serve the same market. Disposable cameras are a well-known example of such a situation (Toktay et al. 2000). In most cases, however, recovered products are seen as distinct and address a specific market segment of price-sensitive customers that choose this product variant in exchange for a price discount. Under these circumstances, it is important to take into account potential demand shifts from new to recovered products, so-called 'demand cannibalization' (Debo et al. 2001).

It is worth emphasizing that product sales, in all their variants, are not the only potential opportunities for recapturing value from reverse product flows. Other options, which are often overlooked, are linked to earlier stages of the original value chain. For example, even products which by themselves are no longer remarketable may still prove valuable on a component level. Recovered components may serve as spare parts, internally or externally, or sometimes as new production input. While in many cases the market value of a product exceeds the value its components and therefore reselling the product as a whole is more profitable a priori than component recovery, exceptions to this seemingly intuitive rule should not be overlooked. Component commonality across product generations, long-term service requirements, and high procurement costs for components late in the life-cycle may shift the balance in favor of the component value.

Moving further back in the original value chain leads to the material content of returned products. Recycling aims at reclaiming these materials. Relatively low raw material prices limit the value potential of this route. Therefore, recycling tends to be profitable only for a few material fractions, notably precious metals. Consequently, it serves as a means for absorbing at least some of the costs of reverse logistics and for avoiding disposal costs, rather than being a driver for initiating new reverse logistics programs. Again, however, exceptions do exist, as illustrated by several nylon-recycling projects in the chemical industry (Realff et al. 2004). In general, such initiatives rely on large scale operations that seek to exploit economics of scale.

In conclusion, reverse product flows may generate value on a product, component, or material level. In general terms, this value may materialize either in the form of cost reductions, by substituting original supply chain inputs, or in the form of revenue increases, by opening new markets. This distinction plays a role in the design of the reverse logistics process, as we discuss in the next section. Despite all of the aforementioned opportunities, it goes without saying that not all reverse product flows are valuable. Some of them represent significant disposal costs in the first place. Therefore, supply control is an important task. However, we repeat that the line between burden and opportunity also depends on a company's creativity and vision to recognize and generate potential markets.

IBM's asset recovery operations illustrate many of the aforementioned reuse options. The priority is on reselling equipment as a whole. To this end, high and mid-range products generally require some reconfiguration, often on the basis of specific customer orders. The products are sold through IBM's regular sales organization as certified remanufactured equipment. A portion of this stream is also sold through business partners and brokers.

For the PC sector, IBM uses a somewhat different channel and does not resell products directly to the customer. Instead, GARS tests these products and then auctions them off in large batches to brokers, who sell them through to specific market segments, for example in Eastern Europe. Other products are donated to schools and charities, sometimes entailing tax credits. Overall, IBM is able to resell some 80% of the PCs returned from the business market.

In a following step, GARS screens the remaining equipment for valuable components. In particular, it supplies spare parts to IBM's own service division. The potential of this so-called 'dismantling' channel relies on the fact that service requirements typically extend well beyond two or three years, the typical duration of a lease contract. In addition, there are few alternative sourcing options for parts towards the end of the service horizon (see also Chapter 9 of this volume). For a detailed discussion of the dismantling channel we refer to Fleischmann et al. (2003). In addition, GARS also sells recovered components to external brokers.

Finally, GARS breaks down the remaining returned equipment into some 50 different material fractions and sells them to specialized recyclers. While precious metals generate some additional revenues, other fractions are sold at a cost.

The above options concern returns from the business market. Returns from the consumer market are less valuable to IBM, given their quantity, quality, and product range. Therefore, IBM often participates in industry-wide solutions for this market sector, as for example in the Dutch ICT take-back program (ICT Milieu 2004). These systems typically rely on material recycling.

8.4   Designing the Reverse Logistics Process

The previous two sections highlighted the sources of and potential market outlets for 'reverse' product flows. The task of reverse logistics is to link these two market interfaces, as illustrated in Figure 8.1. The literature groups the processes that provide this link into a few generic steps (Fleischmann, 2003):

- *Acquisition* (or *collection*) refers to the initial transaction by which a company gains possession of the products;
- *Grading* (or *disposition* or *inspection*) denotes the sorting of the product stream into fractions of different quality and their allocation to different reuse options;

- *Re-processing* includes all transformation processes that prepare products for their future use;
- *Re-distribution* means the delivery to a new market.

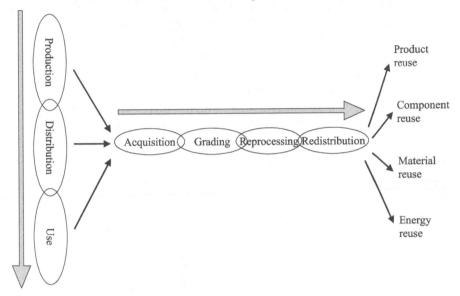

Fig. 8.1. The Reverse Logistics Chain

The collection of these processes forms a supply chain of its own right. Consequently, the individual steps should be coordinated, based on their common underlying goal, namely generating a maximum of value. As in any supply chain, this requires decisions on, among other things, the allocation of processes to different actors, their geographical location and connection through transportation, and the timing of their execution. We address these issues in what follows. Comparing the above 'reverse chain' with conventional supply chains, two processes deserve special attention, namely the acquisition and the grading steps, which differ from conventional sourcing and supply. We devote a separate subsection to each of these processes below. In contrast, the roles of re-processing and re-distribution essentially resemble those of traditional production and distribution operations.

Another aspect that deserves extra emphasis is the fact that the 'reverse chain' is not isolated but, by definition, builds on some preceding 'original' chain. Similarly, additional chains may follow. Figure 8.2 illustrates this view. In some cases, successive chains may literally form a closed-loop that repeats itself. In many other cases the different chains serve different markets. In any case, however, they extend the traditional supply chain framework to a framework that includes multiple use stages. Given their interrelation, supply chain management thinking suggests that the individual chains should be considered as one entity and be coordinated such as to maximize their overall performance. One example that illustrates these interrelations concerns the original product design, which obviously influences all subsequent uses. Through 'design for reuse' or 'design for disassembly' companies explicitly take multiple use cycles into account, in particular by exploiting modularity (Krikke et al. 2004). From a supply chain management perspective, the 'use' stages play a particularly important role since they actually generate the chain's revenues. Managing these stages therefore is a critical lever for coordinating the extended chain as a whole. We return to this issue in our subsequent analysis.

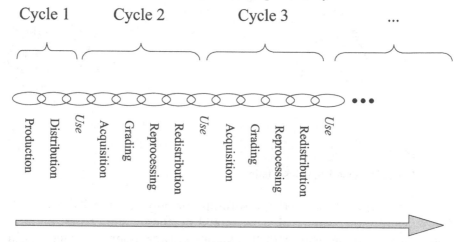

Fig. 8.2. The Extended Supply Chain

8.4.1 Take-Back Strategies

In Section 8.2 we listed sources of reverse product flows. The next question is how companies can use these sources to obtain their desired inputs. Comparing the aforementioned sources with traditional suppliers reveals a number of structural differences. In a traditional buyer-supplier relation, the buyer simply orders the desired quantity at a given price. In a reverse logistics setting, the buyer's choice is often more limited since supply is a derivative of a preceding supply chain cycle (see Figure 8.2). Supply may therefore not be available in the desired quantity or quality. What is more, some transactions may be supply driven rather than demand driven. That is, a supply push partly replaces a demand pull. This relationship is the most obvious for commercial returns and in the case of extended producer responsibilities, which simply oblige companies to take back what customers return. However, even in economically-driven reverse logistics initiatives many companies to date follow a rather reactive approach. While this is a logical choice in some cases, it reflects a lack of awareness in others. In conclusion, the sourcing challenge in reverse logistics is twofold: getting what you want and avoiding what you do not want.

One way to approaching this challenge more proactively is by influencing supply through financial incentives, that is by offering buy-back prices differentiated by product type and quality and dynamically updating them. It is worth noting the particular market mechanism of this setting. Instead of a supplier offering products at a given price, we have a buyer soliciting for products at a given price.

However, the potential of innovative concepts reaches much further. In particular, novel approaches exploit the interrelation between the different phases of the extended supply chain (see Figure 8.2). Rather than losing sight of their products once they reach the customer and then rediscovering them later through reverse logistics, companies may rather monitor the entire underlying process. This opens the way to a conscious trade-off between costs and revenues, and thereby to maximizing the overall value of a product. Note that this approach, which is also known as 'installed-base management' (see van Nunen and Zuidwijk 2004), matches well with the ongoing trend from a physical-product focus to a service focus: selling mobility rather than cars, connectivity rather than mobile phones, and documenting-capabilities rather than copiers. Leasing is a classical implementation of this concept. Service contracts are another example. Selling services whereas keeping the physical products in their own possession, enables companies to optimize the use of these products, by deciding on replacements, maintenance, upgrading, and disposal. Advancing information-technology capabilities further facilitate these decisions by routinely providing rich sets of relevant data regarding, e.g., product wear, usage statistics, and market profiles. As an aside we note that these developments also entail challenging issues regarding security and confidentiality. While to date few if any reverse logistics programs fully exploit these capabilities, we expect them to mark the future development of this field.

Let us return to the case of IBM. As discussed, the majority of the products managed by IBM's asset-recovery organization originates from expiring leases. In principal, these product returns are known in advance, based on the lease portfolio. However, actual return dates and quantities deviate from a simple one-to-one projection, since customers may request contract extensions and, in particular, since they may purchase the product when the lease expires. All in all the actual return process, of both leases and other types, is largely customer driven. Only in a few exceptional cases does GARS actively seek to take back a specific product. Besides customer preferences, actions in IBM's original sales channel are another factor that determines product returns, for example through 'old-for-new' exchanges.

All in all, these different factors result in a fairly stable supply rate to IBM's asset recovery operations, with some slight seasonal fluctuations, driven for example by customers' budget cycles. In general, customers are responsible for shipping returned products to a national IBM return center. From there on, GARS is taking responsibility for further processing.

8.4.2 Grading and Disposition

In contrast with traditional supply, reverse logistics flows, in general, consist of a heterogeneous mix of products of different quality and value. Therefore, the reverse chain typically includes some type of grading and sorting process, which determines the status of the individual products and assigns them to corresponding reuse options. This process is of prime importance as a means of quality control. In addition, its design has a significant impact on the performance of the reverse supply chain and therefore merits specific discussion at this point.

The degree of centralization of the grading and sorting process gives rise to a trade-off. As usual, centralization tends to reduce investment costs by exploiting economies of scale. In the case of the grading process this regards testing-equipment and the required skills to operate it. On the other hand, de-central grading close to the source may reduce transportation costs by separating waste, which ought to be disposed locally, from valuable products, which merit further processing. What is more, de-central grading provides earlier supply information and may thereby speed up the recovery process as a whole. Blackburn et al. (2004) point out that this effect acts, to some extent, in the opposite way of postponement. While in traditional supply chains *delaying* product differentiation creates an option value, *revealing* product differences *earlier* creates value in reverse logistics.

Many of today's reverse logistics programs choose for a centralized grading and sorting process. In line with our argumentation in Section 8.4.2, we see information technology as a factor that may reverse this choice. Remote access to detailed product data reduces the need for physical inspection and corresponding investments and may partly substitute physical flows by information flows.

IBM's asset recovery process involves a two-step grading process. The first step is based solely on nominal product type and model, rather than on individual product identity. GARS selects the types and models that qualify for further use. The selection criteria are dynamically updated, based on market developments. The selected products then undergo detailed individual tests at a central recovery facility.

8.4.3 Location and Network Design

Designing a logistics network for the extended supply chain involves decisions on where to locate the aforementioned transformation processes, notably grading and re-processing, as well as intermediate storage processes, and at what capacity levels. At the same time, corresponding transportation links need to be established. The previous subsections highlight some of the particular issues regarding these decisions. In addition, many of the traditional trade-offs also apply in this particular supply chain context. These include economies of scale both in transportation and in facility investments, consolidation versus responsiveness, and labor cost savings versus transportation.

Given these analogies with conventional supply chain environments, it comes as no surprise that most of the corresponding decision support models in the literature closely follow up on traditional network design models. In particular, many authors have proposed variants of mixed integer linear programming (MILP) facility location models that include 'reverse' supply chain processes (Fleischmann et al. 2003). A few more fundamental extensions include stochastic modeling elements to account for the significant uncertainty that is typical of many 'reverse' supply chains (Realff et al. 2004). However, in many cases it appears that the benefits of these more involved modeling approaches are limited compared, e.g., to simple scenario analyses. This conclusion hinges on the well-known 'robustness' property of network design decisions, in the sense that moderate demand changes do not require a fundamental network re-design. Besides, transportation differences of a few hundred kilometers rarely are a decisive factor on a global scale.

A factor that deserves specific mentioning when it comes to logistics network design for the extended supply chain is potential synergies between different processes. This concerns, in particular, the relation between 'forward' and 'reverse' processes, such as distribution and collection, or original manufacturing and re-processing. For example, combining inbound and outbound transportation may increase vehicle capacity utilization. Similarly, co-locating manufacturing and remanufacturing operations may give rise to economies of scale. On the other hand, separating these processes allows for a more tailored network design, and thus a trade-off has to be made.

In many of today's business examples we observe that companies take a hierarchical approach, in the sense that they give priority to designing the traditional 'forward' supply chain processes and only later fit in reverse logistics processes. Given the interrelations between these sets of processes one may wonder whether this successive approach is in fact appropriate or whether an integrated design of the overall process chain would be superior. In a previous study we have shown that, with respect to the logistics network design, in many cases there is no need to deviate from the common hierarchical approach (Fleischmann et al. 2003). It is again the aforementioned robustness property that allows one to decompose the overall network design into two separate parts. Exceptions include those cases, in which recovered product content substitutes 'virgin' supply and both streams differ significantly in their cost structure. This applies, for example, to the substitution of pulp wood from Scandinavia by recycled paper from Western Europe in paper production.

The above general considerations and trade-offs are also reflected in IBM's asset recovery network. Besides the aforementioned national return centers where returns are selected and consolidated, GARS operates two major recovery facilities in the EMEA region, each for a specific product range. These are located in Montpellier, France, for the server equipment and in Mainz, Germany, for all other product ranges. These facilities host the final grading operations, the actual remanufacturing and subsequent storage, as well as component disassembly and material separation. Currently, PC operations are subcontracted to a service provider, while other operations are carried out in-house. To achieve economies of scale, all transportation operations are subcontracted to a single provider that is also responsible for the 'forward' distribution shipments.

The network structure in the other geographical regions, i.e. America and Asia Pacific, closely resembles that in EMEA. Specifically, GARS operates two product-specific facilities in the U.S.A., in addition to central facilities in Canada and Brazil, as well as in Japan and Australia. The number of facilities in each region, and thus the degree of centralization of the recovery operations, is a major strategic choice. In contrast, the exact location of these facilities is largely historically determined, based on available competencies and infrastructure. This illustrates once more the common hierarchical network design approach discussed above.

8.5   Inventory Management and Value of Information

Another important element of the supply chain design, besides the geographical location of the various processes, is their inter-temporal coordination. This relates to the location of inventory buffers, which decouple the individual processing steps. Traditional supply chain management commonly distinguishes inventories according to their supply chain function, such as cycle stock, seasonal stock, and safety stock. All of these functions also play a role in the extended supply chain. Moreover, inventories assume an additional role in this context, which is driven by the mismatch between exogenous supply and demand. Since, in general, customers do not return products exactly at the moment that these can be resold, companies build up inventories of re-marketable products, which we denote as 'opportunity stock'. The effect is similar to that of forward buying in response to a temporary price discount.

An important choice in any supply chain design concerns the location of the customer-order decoupling point, i.e. the borderline between make-to-stock (MTS) and make-to-order (MTO) processes. In the extended supply chain, each usage cycle contains an additional such decoupling point on the supply side (see Figure 8.3). This point indicates how far in the process chain a returned product moves upon its arrival. Analogous with traditional terminology one might denote the processes after and before this point as 'make-*from*-stock' and 'make-*from*-supply' processes, respectively. Needless to say, both decoupling points may coincide in a single inventory buffer.

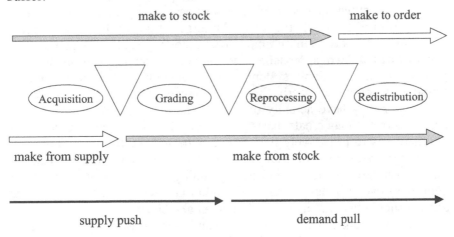

Fig. 8.3. Inventory Buffers in the Reverse Logistics Chain

A related, but not identical, supply chain characteristic concerns the border between supply-push and demand-pull processes. In particular, it is important to decide whether the re-processing stage, which typically represents the main value-adding activity of the extended chain, is to be push or pull-driven. In the former case, one processes returned products as they become available, whereas in the latter case one postpones value-adding activities until demand materializes. In a study on IBM's component-dismantling operation we highlighted that the appropriate choice depends primarily on how certain one is of future demand for the product in question (Fleischmann et al. 2003). In case of a serious risk of not finding a demand, and thus of wasting the reprocessing expenses, it is advisable to postpone any costly re-processing until more demand information becomes available. In all other cases, postponing the re-processing operation comes down to trading higher safety stock levels against lower holding costs per unit, which in sum leads to slight inventory cost savings at best.

The management of seasonal stocks and cycle stocks in the extended chain does not appear to differ essentially from traditional environments. The literature provides several variants of economic-order-quantity (EOQ) models for lot-sizing decisions in product recovery operations (Minner and Lindner 2003). In contrast, choosing appropriate levels of safety stock and opportunity stock is more challenging. A significant body of literature addresses this issue (see, e.g., van der Laan et al. 2003). What complicates the matter in the first place is the additional uncertainty on the supply side of the extended chain. Higher overall uncertainty typically implies the need for higher safety stock buffers. A second complicating factor concerns the fact that returned product content and new products and components often serve as substitutes, as for example in IBM's service operations. In this situation, one needs to coordinate multiple alternative supply sources with different characteristics in terms of cost, reliability, and lead times, in such a way as to minimize overall costs (see also Chapter 9). One can distinguish two approaches for integrating market returns into the planning of such a supply system. Most commonly in current practice we found a conservative, reactive approach, which only takes returns into account after they have actually occurred. The downside of this 'safe' approach is that it may create excessive inventories of unneeded returns. The alternative is to proactively incorporate expected future return flows into the current planning, for example when ordering new components. We have illustrated that such a proactive planning can significantly reduce inventories, even though it requires additional safety buffers to protect against supply uncertainty (Fleischmann et al. 2003).

Inventory management critically depends on the available information about future supply and demand, and thus in particular on forecasting. Just as in traditional supply chains, managing the extended chain requires projections of future demand. Expert assessments and statistical tools provide a basis for such estimates. What is more particular is the forecasting requirement on the supply side of the reverse logistics chain. In the literature, different methods have been proposed for estimating future product returns, which form the basic resource of the extended chain (Toktay et al. 2003). Simple methods treat the return flow as an autonomous process and apply the same statistical techniques as in demand forecasting. More advanced methods explicitly capture returns as a consequence of a previous supply chain cycle (see Figure 8.2). From this perspective, the key is to estimate the time a product spends in the market. Since this approach requires demand information from the previous supply chain cycle it is particularly appealing to OEMs that collect and recover their own products. Yet, even if historical demand information is available it may be non-trivial to determine the actual time that a product spent in the market. While the sales history of a high-end product in the business market may be well documented, this is not the case, in general, for commodities such as PCs, disposable cameras, or even softdrink bottles. However, as discussed previously, advances in information technology are about to change this picture. The ever more widespread and cheaper availability of digital storage devices, such as RFID-tags, provides the basis for tracking detailed product data even for simple commodities. Heineken's 'Chip-in-crate' project nicely illustrates this development. In this pilot project, the Dutch brewer equipped a set of its reusable beer crates with an electronic chip that is read whenever the crate passes through the bottling line (van Dalen et al. 2004).

As discussed in Section 8.4.1, the impact of this new wealth of information reaches much further than providing a basis for more reliable forecasting. For example, it may replace forecasts by real-time actual data. Moreover, it allows companies to realize an active acquisition management, that is to manage the supply side of the extended chain rather than to accept it as purely externally given.

These quickly expanding possibilities raise the question which type of information is the most critical for enhancing the extended supply chain and how to quantify its actual benefits. A stream of literature on the 'value of information' focuses mainly on inventory cost savings through the reduction of uncertainty (Toktay et al. 2003). Yet it appears that other benefits of advanced product information are even much larger. In particular, information helps identify potential supply and demand and thereby enables valuable transactions that otherwise would not have been realized at all. Pricing decisions are another issue that is closely related to this type of information. Finally, supply and demand information is key to supply chain design decisions such as capacity investments. In our opinion, a systematic and detailed analysis of the factors that determine this broader 'value of information' is one of the primary current research mandates in this field.

In order to position IBM's asset recovery processes in the above inventory management framework we need to distinguish several channels. As discussed in Section 8.4.1, the supply of returned products is essentially customer driven. For the PC product range this supply push extends all the way to the re-processing operation, i.e. the testing of the returned PCs. In fact, even the re-selling by means of auctioning can be characterized as a push operation. In this way, safety stocks and opportunity stocks can be limited. Since return rates turn out fairly stable, seasonal stocks do not play an important role either. Inventory occurs mainly as cycle stock at the end of the operation while waiting for a sufficient auctioning batch to accumulate. For higher product ranges, the push-pull border lies further upstream in the process. For these products, the required reconfiguration depends on specific customer requests and is therefore carried out in a make-to-order fashion in most cases. Consequently, the major inventory buffer contains preliminarily tested equipment awaiting reconfiguration. Only for a limited product range, full testing and reconfiguration are carried out immediately after receipt, in order to have the products available for fast re-sale. In both of these channels, IBM uses supply forecasts mainly on a medium-term aggregated level to adjust capacities. In contrast, short-term forecasting turns out to be difficult, even for leased equipment (see Section 8.4.1).

Also IBM's supply of service parts from dismantled machines is push-driven. Available parts which match projected future demand are moved through their specific recovery process and are then stored ready-to-use until actually being deployed in the service network. Again, supply forecasts are mainly used in the long-term planning, namely the expected contribution of different supply sources during the product lifecycle. Besides for cost calculations, this information is important, for example, for deciding on the size of the final production order at the time that production of new parts is phased out.

8.6 Conclusions

Product flows in today's supply chains do not end once they have reached the customer. Many products lead a second and even third or fourth life after having accomplished their original task at their first customer – or after this customer changed his mind and returned them. Initially, many of these additional products flows were driven by ecological arguments, namely waste reduction, and by customer service obligations. Consequently, many have seen product returns as a cost factor in the first place. In the meantime however, companies have started recognizing the potential value of these flows. Instead of a single time, a product may generate revenues multiple times, possibly in different markets.

Capturing this value requires a broadening of the supply chain perspective. This broader view includes new processes, such as the collection of products from the market and the grading of these products according to their quality and future value. More importantly, it includes multiple interrelated usage cycles, linked by specific market interfaces. Coordinating the successive product uses is key to maximizing the value generated.

To date, many companies deal with product returns in a purely reactive manner. While in some cases it does, indeed, make good sense to give unlimited priority to the initial product market this strategy is shortsighted in many other cases. Maximizing a product's lifetime value requires a more proactive attitude. In particular, it requires a good understanding of the interrelations between different phases of the product lifecycle. Market incentives can then help assign the product to its most valuable use at each time.

Information technology is a key enabler of this integral approach. Timely availability of detailed product, process, and market data allows companies to manage the corresponding processes in a conscious way. The current realization of extended supply chains is still in its early stages. Their potential is huge.

References

Blackburn, J.D., Guide, Jr., V.D.R., Souza, G.C., Van Wassenhove, L.N. (2004) Reverse Supply Chains for Commercial Returns, *California Management Review* 46(2): 6–22

de Brito, M. P., Dekker, R. (2003) A Framework for Reverse Logistics, in: Dekker, R. et al. (eds.), *Reverse Logistics: Quantitative Models for Closed-Loop Supply Chains*, Chapter 1, Springer Verlag, Berlin

Debo, L.G., Toktay, L.B.,Van Wassenhove, L.N. (2001) Market segmentation and product technology selection for remanufacturable products, Working Paper, INSEAD, Fontainebleau, France

Dekker, R., Fleischmann, M., Inderfurth, K., Van Wassenhove, L.N. (eds.) (2003) *Reverse Logistics: Quantitative Models for Closed-Loop Supply Chains*, Springer-Verlag, Berlin

European Commission (2004) Waste Electrical and Electronic Equipment. Retrieved on Sept. 10, 2004 from http://europa.eu.int/comm/environment/waste/weee_index.htm

Flapper, S.D.P., van Nunen, J.A.E.E., Van Wassenhove, L.N. (eds.) (2004) *Managing Closed-Loop Supply Chains*, Springer Verlag, Berlin.

Fleischmann, M. (2001) *Quantitative Models for Reverse Logistics*, Springer Verlag, Berlin

Fleischmann, M. (2003) Reverse logistics network structures and design, in: Guide, Jr. V.D.R., Van Wassenhove, L.N. (eds.), *Business Aspects of Closed-Loop Supply Chains*. Carnegie Mellon University Press, Pittsburgh, PA

Fleischmann, M., van Nunen, J.A.E.E., Gräve, B. (2003) Integrating closed-loop supply chains and spare-parts management at IBM, *Interfaces* 33(6): 44–56

Fleischmann, M., Bloemhof-Ruwaard, J.M., Beullens, P., Dekker, R. (2003) Reverse Logistics Network Design, in: Dekker, R. et al. (eds.), *Reverse Logistics: Quantitative Models for Closed-Loop Supply Chains*, Chapter 4. Springer Verlag, Berlin.

Guide, Jr., V. D. R. and Van Wassenhove, L.N. (eds.) (2003) *Business Aspects of Closed-Loop Supply Chains*, Carnegie Mellon University Press, Pittsburgh, PA

ICT Milieu (2004) The ICT take-back system. Retrieved on September 10, 2004 from http://www.ictmilieu.nl

Krikke, H., le Blanc, I. and S. van de Velde (2004) Product modularity and the design of closed-loop supply chains, *California Management Review* 46(2):23–39

Minner, S., Lindner, G. (2003) Lot-Sizing Decisions in Product Recovery Management, in: Dekker, R. et al. (eds.), *Reverse Logistics: Quantitative Models for Closed-Loop Supply Chains*, Chapter 7. Springer Verlag, Berlin.

Realff, M.J., Ammons, J.C., Newton, D.J. (2004) Robust reverse production system design for carpet recycling, *IIE Transactions* (forthcoming)

Rogers, D. S., Tibben-Lembke, R.S. (1999) *Going Backwards: Reverse Logistics Trends and Practices*. Reverse Logistics Executive Council, Pittsburgh, PA

Toktay, L.B., Wein, L.M., Zenios, S.A. (2000) Inventory management of remanufacturable products, *Management Science* 46(11): 1412–1426

Toktay, L.B., van der Laan, E.A., de Brito, M.P. (2003). Managing Product Returns: The Role of Forecasting, in: Dekker, R. et al. (eds.), *Reverse Logistics: Quantitative Models for Closed-Loop Supply Chains*, Chapter 3, Springer Verlag, Berlin.

Tsay, A.A., Nahmias, S., Agrawal, N. (1998) Modeling Supply Chain Contracts: A Review, in: Tayur, S. et al. (eds.) *Quantitative Models for Supply Chain Management*, Kluwer Academic Publishers, Boston, MA

van der Laan, E.A., Kiesmüller, G., Kuik, R., Vlachos, D., Dekker, R. (2003). Stochastic Inventory Control for Product Recovery, in: Dekker, R. et al. (eds.), *Reverse Logistics: Quantitative Models for Closed-Loop Supply Chains*, Chapter 8, Springer Verlag, Berlin

van Dalen, J., van Nunen, J.A.E.E., and C. Wilens (2004). The Chip in Crate: The Heineken Case, in: Flapper, S.D.P. et al. (eds.), *Managing Closed-Loop Supply Chains*, Springer Verlag, Berlin

van Nunen, J.A.E.E., Zuidwijk, R.A. (2004) E-Enabled Closed-Loop Supply Chains, *California Management Review* 46(2): 40–54

9   Service Parts Logistics Management

Michel W.F.M. Draper and Alex E.D. Suanet

9.1   Introduction to Service Parts Logistics

9.1.1   Overview of this chapter

In this chapter we give an overview of recent Service Parts Logistics Management developments within IBM. In particular we will make a comparison with traditional Supply Chain Management and describe the specifics in the Service Logistics Supply Chain. While the concepts we describe here are applied within IBM's Service Logistics worldwide, specific examples will be given for the European situation as supported by the IBM Service Logistics organization in the Netherlands. Because of rapid changes in the Service Logistics environment we also include a vision about developments in the foreseeable future.

In the first section we will discuss similarities and differences between the Service Logistics Management process and the more traditional Supply Chain Management process. We will indicate the Service Logistics Management process can be seen as a specific kind of a Supply Chain Management process with some additional specific elements and complexities.

In the second section we will discuss the Service Logistics Management process in more detail. The most important sub-processes will be described shortly. A distinction will be made between the preparing and planning processes and more operational oriented processes.

In the third section system solution developments will be mentioned briefly. This will not be an extensive list of systems but more an illustrative section indicating with what kind of system environment the most important processes are supported.

In the fourth section we will briefly describe expected future developments in the Service Logistics Management process.

Finally, at the end of this chapter a section Further Readings is included which can be used to find more details.

9.1.2 Service Parts Logistics

Service Parts Logistics, often referred to as Spare Parts Logistics, is defined as the flow of service parts across the entire supply chain from manufacturers and vendors to the distribution network and via this network to Client Engineers who use these parts for repairing machines. Service Parts Logistics has already a history of many decades. Nowadays it becomes more and more popular. This mainly due to the fact that the business is shifting from a product oriented market to a service oriented market. The clients of today require and demand higher service. This is a consequence of the growing integration of business and information technology. Today, many businesses face a mission critical information technology component. This is directly related to the Service Logistics Management process which has become more and more important.

The Client Engineers being part of the Service Delivery organization use the service parts primarily for machine repair of machines installed in the field. Usually these machines are covered by a service maintenance agreement. In addition to this, a limited portion of parts required, is used for machine repair of machines that are within their warranty period.

Service parts are also used on a per call basis, meaning that no service is provided based on a maintenance agreement but on a per call basis. Usually this means that a machine is repaired based on time and material. The time is related to the required Client Engineer labor time to fix the machine and the material is related to the service parts and supplies used during repair. A relatively small percentage of parts are sold directly to dealers and clients. This means that the part is not used by IBM's own service delivery organization. The parts can be sold to external parties that use these parts for machine repair services. A fairly new way is that service parts are sold directly to clients using an on-line shop. In the shop the client can search the service part needed and place an order (for further information see http://www.ibm.com).

In addition to traditional Supply Chain Management processes the Service Parts Logistics Management process deals also with matters like probabilistic demand, longer lifetimes of the parts in the supply chain, possibilities of substitutions (and as a consequence additional complexity) and Reverse Logistics.

9.1.3 Probabilistic Demand

In the Service Parts Logistics business the part demand is the result of a probabilistic process of part failures. Most service parts have a sporadic and uncertain demand pattern. This makes it almost impossible to use forecasting techniques developed for traditional Supply Chain Management processes. Special forecasting methods are or need to be developed to handle this sporadic demand. In addition to historical demand other data is used as well, like machine install base information and marketing prognoses. Calculating forecasts in Service Parts Logistics is very difficult, while it is one of the most important elements in decision making (stock planning). For more details we refer to Silver, Pyke and Peterson (1999), Chapter 4.

9.1.4   Lifetime

The lifetime of the parts within the chain can be much longer in Service Logistics. Parts flow through the chain and are stored within the distribution network until the moment they are required. This may take several years or it might be even the case that they are never used at all. This is due to the fact that the probability a specific part is required is usually very low, but if it is required it need to be delivered on a very short term (within hours) to the Client Engineer. See Sherbrooke (1992).

9.1.4.1.   Substitutions

The lifetime of the service parts are closely related to the lifetime of the products in which they are used. Generally product lifetimes vary from a few years up to the range of 15 to 25 years. Because of the continuous product improvements the phenomena known as substitutions is introduced. In most cases substitutions are defined as functional equal parts but with different technical characteristics. Substitutions exist because of the continuous technical improvements made to products and the parts they contain.

There are different types of substitutions. Examples are fully interchangeable substitutions and forward substitutions. A fully interchangeable substitution means that the substituted (old) part and the substitute (new) part can be put in any machine without a problem. In case of a forward substitution the substituted part can be exchanged by the substitute part, but not the other way around. Nowadays a special kind of substitutions is used as well. These are called matrix substitutions. The parts have equal form, fit and function but differ in price. In addition the prices change continuously which makes control of the process much more complicated. In the Service Logistics supply chain multiple versions of functionally the same part are stored. In the forecasting, planning, ordering and delivery processes these substitutions relationships need to be taken into account, which is not the case in a traditional Supply Chain Management process.

9.1.4.2. Reverse Logistics

Today the reverse logistics is getting a more important part in Supply Chain Management. In Service Parts Logistics it has always been very important to control the flow of returned parts. Service parts that are returned, have the qualification 'good' or 'defective'. They may have the indication 'good' because of several reasons. One of them is that they are needed by a Client Engineer together with another part for diagnosis. The part necessary for the machine repair is used while the other is returned. The temporary use of parts by a Client Engineer for diagnosis is called the 'On-Loan' process. The On-Loan process is applied in EMEA[8] only. While the part is on loan it still belongs to the Service Logistics organization. When it is used for machine repair it is transferred to the Service Delivery organization. The Service Delivery organization in EMEA is organized at a country level.

Parts that are defective follow a reverse flow as well. Many of these parts can be repaired and more and more specialized companies provide specialized repair services to the market. Therefore repairable defective parts flow to specialized repair vendors. After these parts are repaired they flow back into the logistics supply chain at a central or local level.

In addition to this the IBM reverse logistics process may also use parts flowing out of the dismantling process. Complete machines that return from the field, for example due to an End-Of-Lease situation, often contain useful and valuable parts. These parts are identified and when predefined conditions are satisfied, flow back into the available parts inventory. See Fleischmann et al. (2003).

9.2   The Service Logistics Management Process

At a high level the Service Logistics Management process can be decomposed into processes as illustrated in Figure 9.1, where the rectangles indicate the processes and the arrows the most important relationships between these processes.

---

[8] EMEA means Europe, Middle East and Africa.

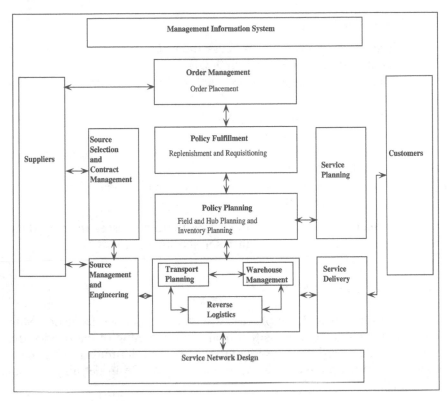

Fig. 9.1. The Service Logistics Process

The supply chain starts at the supplier side. In the end the suppliers such as manufacturers and vendors provide the service parts required for service maintenance purposes for products installed at client sites. The parts flow from the supplier side to the client side being managed via this process. The sub-processes can be grouped into preparing and planning processes and operational processes.

Preparing and planning processes:

- The Service Planning process, defines what service is required (see section 9.2.1).
- The Service Network Design process defines what service network is needed to realize the required service (see section 9.2.2).
- The Policy Planning process defines which inventory levels are needed (see section 9.2.3).
- The Source Selection and Contract Management process defines which sources are needed (see section 9.2.4).

Operational processes:

- The Policy Fulfillment process monitors the inventory and constraint situations (see section 9.2.5).
- The Order Management process monitors the orders (see section 9.2.6).
- The Source Management and Engineering process monitor the supply sources (see section 9.2.7).
- The Service Delivery process monitors and manages client requests (see section 9.2.8).
- The Management Information System via which all reporting is done (see section 9.2.9).

In subsequent sections we will describe these processes in more detail and how they fit into the maintenance service strategy.

9.2.1   Service Planning

One of the fundamental sub-processes within the Service Logistics Management process is the process that deals with planning of the service that needs to be delivered. Service Planning deals with defining service levels related to various products. Elements that play a role in defining this are for example:

- Market expectations (client needs and expectations).
- The type of client agreement (service maintenance contract, warranty services, per call service and others).
- The type of product (categorized into divisions and sub-divisions).
- The importance of the service part (categorized into service part vitality classes).
- The inventory budget.
- The geographical area (high density business area, moderate business area or remote area).

The transformation and improvements IBM is making in this process have been going on for years. There are two forces in effect here. The first is that from an overall cost perspective the inventory needs to be minimized. The second is that the clients have more and more needs and want value for their money. These forces resulted in a shift of the concept of service itself. Till a few years ago, service was expressed as the availability of the demanded parts directly from shelf. Today this is not sufficient anymore. In order to meet the new targets, of course the parts need to be available but also within in a defined time frame. This has far-reaching consequences for business processes and the supporting system environment. In the part planning as well as in the part delivery processes initiatives are ongoing to be in line with this new strategy. Studies in this area investigate, for example, making use of the client maintenance contract information and the geographical positions where machines are installed.

Within the service planning process a translation of service requirements into parameters and criteria is made to be used in the stock planning process. Service parameters used in the internal processes are expressed in terms of a Parts Availability Level (PAL). The service levels are usually defined for specific sets of products in combination with a geographical area to which they apply.

While the Parts Availability Level has been used over the last few decades as the primary service parameter, this concept is due for change. More and more it is being replaced by the Parts Delivery Time (PDT) parameter. As such the transformation to provide differentiated service fits in the IBM 'e-business on demand' strategy. It is an important change in the current Supply Chain Management process. See also Simchi-Levi et al. (2004).

This strategy focuses on the fact that within the market different needs exist and a different response and different service are required. In the Service Planning process the client needs and expectations will be translated into PDT parameters used in the internal process. The PDT expresses the delivery time of a service part from the moment a service part request has been accepted and validated till the moment it is delivered at the point of final destination. PDT depends on elements like the client service needs, type of product and the geographical area.

9.2.2 Service Network Design

Another fundamental element within the Service Logistics Management process is the design of the infrastructure. See also Drezner and Hamacher (2004). The infrastructure or network design refers to the positioning of warehouses called 'stock locations' and the arrangement of transportation connections between these warehouses, spanning a geographical area, such that from a logistics point of view the service requirements can be met.

Worldwide Network

The global service logistics network is divided into several geographical networks as shown in Figure 9.2. The geographical networks each have a central warehouse, also called a hub. The hubs buy and receive parts from many manufacturers and vendors. This is illustrated by the nodes labeled 1, 2, a, b etc.

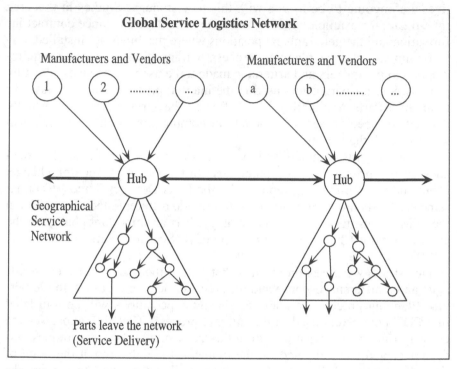

Fig. 9.2. IBM's Global Service Logistics Network

The current worldwide Service Logistics network of IBM has four major hubs:

- Amsterdam, in the Netherlands covering the EMEA geography.
- Mechanicsburg (Pennsylvania) covering the Americas and Asia Pacific geographies.
- Singapore covering the Asia Pacific geography (except Japan) for a limited set of parts.
- Tokyo covering Japan (also for a limited set of parts).

The geographical networks are mainly connected via the hubs. Manufacturers and vendors, (usually) located in the same geography as the hub, deliver the ordered parts to the hubs. From the hubs the parts are shipped into the geographical networks or transshipped to other hubs.

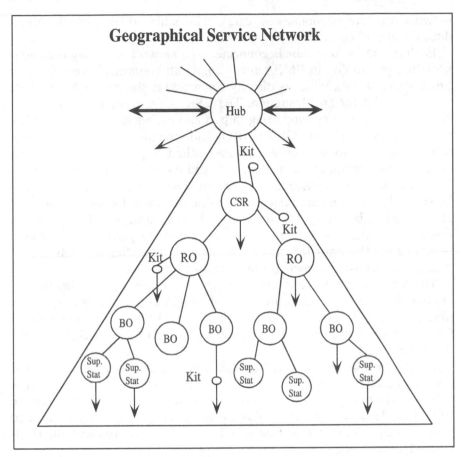

Fig. 9.3. Geographic Service Network

Geographical Network

The physical structure of the geographical networks has been developed based on historical decisions and local circumstances. The optimization of the physical network structure is an ongoing process. With specialized optimization algorithms the layout of the network can be determined. The creation of the network structure like warehouse position determination and the creation of transportation connections between warehouses and clients, is a fundamental element in the Service Logistics Management process. IBM's strategy in this respect is to form partnerships with Logistics Service Providers. The focus of IBM in the process is the translation of the underlying IBM-Client business requirements. The focus of the Service Logistics Provider is the process of managing the physical infrastructure like the warehouses and transportation connections. Due to the size of the network structure, economies of scale can be achieved here, which will reduce the overall costs.

Each of IBM's four hubs is connected to a network spanning the corresponding geography. In EMEA for example all countries have a Country Stock Room (CSR), which is directly connected to the hub in Amsterdam. See Figure 9.3 for an illustration. The CSR is the most important warehouse within a country supporting import and export requirements. CSR's are connected to Region Offices (RO) and Branch Offices (BO), sometimes connected to smaller warehouses called Support Stations. In this context a connection means that an arranged transportation connection exists between these warehouses. To support specific clients, parts may also be stored in kits. Kits are related to a particular stock location. Conceptually these can be seen as small dedicated stock locations. The arrows in Figure 9.3 indicate some possible flows of service parts. Arrows downward indicate that parts flow out of the network to the Service Delivery organization that uses the parts for maintenance purposes.

The network is a layered structure. The layers are called echelons. Including the hub the echelon hierarchy in EMEA is five levels deep, spanning more than 50 countries. It includes about 230 warehouses (1 hub, 35 CSR's, 26 Region Offices, 137 Branch Offices, 31 Support Stations) and 2,000 Kits

Having an efficient logistics network is one of the key elements used to control the cost of the entire Service Logistics process. The warehouse network and transportation facilities are outsourced to Logistics Service Providers. This is a process followed and managed by IBM. Dedicated projects are run to select the best suited networks compared with the IBM-Client business requirements.

Same Day and Next Day Networks

In order to keep up with new business requirements, IBM moves to a differentiated service network structure. This has lead to the development of two different cooperative network structures. The first network structure is aiming for clients, which demand mission critical service and need a fast response. This network is called the 'Same Day' service network. The Same Day network consists of a relatively high density structure of warehouses. Transportation to the point where the service part is required only takes a few hours. An important factor of influence in deciding the quality of such a network are the client machine install positions, the time it takes to deliver the demanded service parts to client sites and the expected frequency of transport connections.

The other network is providing a more standardized service to clients for less critical situations. This network is called 'Next Day' service network. Alternatively it is referred to as the 'Direct Client Shipment' network, since ideally service parts are shipped from the hub 'overnight' directly to (or close to) the client site. The majority of Europe can be reached overnight from the hub in Amsterdam. For situations where this is not really possible a limited number of local Next Day hubs support the central warehouse in this. Next Day stock locations are combined with a series of so called 'Pick Up' and 'Drop Off' points. The Pick Up and Drop Off points are small locations used by the Client Engineers to pick up the service parts that they need.

Important decision factors in the construction of the Next Day network are the expected frequency of shipments to the Pick Up and Drop Off points and costs of transportation from the hub and Next Day stock locations. The Next Day network structure contributes to the reduction of the amount of service parts that need to be stocked throughout the entire network, and to the reduction of transportation costs due to consolidated transport and cheaper transportation facilities.

The Same Day and Next Day networks are interwoven as well. This because several physical locations might perform multiple functions, i.e. they can perform Same Day part deliveries, Next Day part deliveries and also replenishment tasks.

Kits

In addition to the Same Day and Next Day networks, service parts might also be stocked at client sites for specific contracts and possible urgent service parts demands. These stock facilities are called 'kits'. Currently kits are usually available only for specific clients and are typically used where the base network is unable to provide the required service.

Commodity Logistics Centers

To reduce the network inventory and make the whole Service Logistics process more efficient, recently the concept of a Commodity Logistics Center (CLC) has been introduced. A CLC is a specialized repair vendor, which provides service parts belonging to a specific group of parts (the commodity). Part demands are not fulfilled from the Service Network but from the CLC. The CLC manages the part supply and fulfills the demand. The CLC does not only deliver to the IBM network but for other parties as well. The selected group of parts and the economies of scale make it possible for the CLC to deliver the demanded parts in a more efficient manner than a single company can do. See also Cachon (2004), for a discussion on Vendor Managed Inventories.

9.2.3 Policy Planning

The geographical networks form the basis for processes executed at a geographical level. Differences between geographies exist, for example, due to differences in markets, culture, business segments and management responsibilities.

In the Policy Planning process, service parts stock levels are determined for the stock locations in the network based on inventory, service and cost targets. Relationships between these three elements are crucial in the management of the Service Logistics process. Because of the importance of the parts inventory value, focus within the Policy Planning process is on budget and target setting as well as on the actual determination of the part stock levels. The first activity is the main objective of the Inventory Management sub-process. The latter are the main objectives of the Hub Planning and Demand Planning sub-processes (sometimes referred to as the Stock Planning process).

In the Inventory Management process, decisions are made on how to allocate the total inventory per geography and product group (or division). On one hand because of the continuous use of parts for maintenance purposes, parts flow out of the service network. On the other hand parts are ordered, received and replenished and flow into the service network. It is the primary task of the Inventory Management process to balance these flows such that the overall inventory targets are met.

In the Hub Planning process the safety stock levels for the central location are determined and set. Safety stock levels are used to cover for unforeseen situations during the supply lead time.

The Demand Planning process focuses on the determination of the service part stock levels at the stock locations in the service network. An important element in the calculation of these stock levels is the life cycle phase of the products that contain these parts. In general a distinction is made between three phases: the Early-Life phase, the Mid-Life phase and the End-of-Life phase. In each of these phases different processes, for forecasting and planning techniques have been developed and are subject to continuous improvement.

Early-Life Phase

IBM will offer service as soon as a product is introduced in the market. Especially in high-end product introductions, high expectations exist. This is why the Service Parts Logistics organization is involved in an early stage of such an introduction.

At this stage there is no historical information available. The data available are sales forecasts, estimated parts failure rates and Service Parts

Lists. This data has the tendency to change frequently, mainly due to new insights. In this life phase specialized systems have been developed and implemented to create forecasts and plans. See Draper and Suanet (2002) and Silver, Pyke and Peterson (1999), section 4.9.

The forecast is based on the sales plans for marketing and the estimated failure rates of the parts in the product. The sales plans are not detailed enough to make an accurate planning. In order to create more detailed forecasts, the historical information of comparable products and the new product planner expertise is used to distribute the volumes as indicated in the sales plan across the network. The new product planner also estimates the growth rate of the new product introduction.

A new product will not always consist of new part numbers only. Usually it also contains already existing part numbers. The new product planner can therefore decide to use the current planning levels as a base for the initial planning of existing parts. The initial planning is calculated based on the initial forecast, the current planning and the service targets. The new product planner can calculate different service scenarios to calculate the effect on the costs.

Mid-Life Phase

When the product matures, it enters the Mid-Life phase. In this phase the policy planning techniques used differ from the Early-Life phase. The most important change is that historical service part demand information is used to calculate an aggregated time phased forecast. Also the effects of substitutions and reallocations of inventory are taken into account in the forecast and planning algorithms.

The aggregated forecast used in the central planning process is based on the net outflow of parts in the entire network. Conceptually the net outflow should be determined by matching the good part returns with the original part demands. This is done in supporting administrative processes.

To calculate the aggregated forecast, a mixture of specific forecast techniques is used. Different techniques are used to calculate the same part number forecast as accurately as possible. Techniques used are, for example, moving average (for 3, 6, 9 and 12 months), exponential smoothing, linear regression, adaptive smoothing and some other specific statistical techniques. Part forecast errors are calculated internally and from the calculating methods a best performing selection is made on a part number level. Finally a blend of the selected best performing methods' forecasts is calculated.

In the central forecasting and planning process, part substitutions are handled via a transformation mechanism. Simply stated, part substitutions form a chain of so-called Engineering Change (EC) levels. Demands on all levels are used as a history of demands for the part with the highest EC level. The translation back to any lower EC levels is made in the operational process during execution.

It is almost impossible and certainly not practical to set a service target for each individual part at each stock location. By doing so, it would mean that around 25 million settings must be defined for IBM EMEA. In order to manage these large amounts, parts are categorized and for each category a service target can be defined. The main criteria used to define these categories are the product ranges, the importance of the part (called part vitality), the part price and the forecast of the expected part demand. Examples of product ranges are z-series, i-series, p-series and x-series[9]. These categories are called product sets. A planner can control the service criteria for each product set at each stock location. It is also possible to create specialized product sets for a specific stock location.

The algorithm used to set planning levels takes into account many elements of the supply chain. For example the geographical network, the forecast, substitutions and current stocking levels. The optimization is done bottom-up, this means that first the stock locations in the lowest echelons are optimized and then the higher echelons stock locations. The demand, which can't be fulfilled at the lowest locations, is passed up to a higher echelon related stock location.

A big difference from the algorithms traditionally used in Supply Chain Management processes is the way the service targets are used. In the Service Logistics process the optimization is done at a product set level, rather then at a part level. This means that the service targets are satisfied for the whole set of parts contained in these products. Based on the demand probability and costs stock levels are calculated automatically in the optimization process such that the required service targets are satisfied. Beside automatic determination of the stock levels the planners can set these levels manually as well.

The field forecast and planning process has focus on the flow of parts going out of the network, i.e. the parts that are handed over to the Service Delivery organization.

End-of-Life Phase

The End-of-Life phase is defined as the period of time between End-of-Manufacturing and End-of-Service. This is a difference between the Ser-

---

[9]   Old names used to indicate these product ranges were mainframe systems (z-series), AS400 systems (i-series), RISC 6000 systems (p-series) and the PC systems (x-series).

vice Parts Logistics process and the traditional Supply Chain Management process. Depending on the part this can be as early as three months after introducing the part (typical for PC parts) or maybe after several years. The length of the End-of-Life phase also depends on the type of product, but is generally assumed to be 7 years.

The Service Parts Logistics organization is responsible for service for the whole lifecycle including the End-of-Life phase. In order to do this a 'last time buy' is made from the manufacturer or vendor. The order quantity mainly depends on the demand forecast for the End-of-Life phase. Therefore much effort is put into this process. Specialized algorithms take into account the historical demand pattern, the projected decline in the machine install base and the required service to be provided in time. See Teunter and Klein Haneveld (2003).

Next to this, refurbishment, reuse of old parts and current stocking levels are taken into account. This information combined in systems designed and implemented for this process, results in a last-time-buy order. The parts are stocked across the service network to guarantee the highest possible service.

9.2.4 Source Selection and Contract Management

The Source Selection and Contact Management processes deals with the identification, selection and contract management with sources of supply. Both processes belong to the Procurement process and are executed under responsibility of the of the IBM Procurement organization. The Procurement organization is responsible for the commercial relations with part vendors. For new-buy sources Procurement cooperates with manufacturers. For repair sources Procurement cooperates with Source Management and Engineering.

In the Source Selection process, the sources of supply which can provide the required parts are defined. Sources of supply can be IBM manufacturing sources as well as vendors from which IBM purchases the service parts. In case of external suppliers, negotiations between IBM and the suppliers are performed that lead to an agreement. Once the agreement is made, the source and important characteristics of it are added to a source menu. The continuous maintenance process of the source/contract menu is called the Contract Management process. When available in the source menu, the potential supplier is made visible for the planning process. Relevant elements of information which are necessary in the planning process include the part number provided by the source, the lead time, the price, the start and end dates of the agreement, the minimal order quantity and shipment costs.

Criteria that are setup are used to manage a work queue for Procurement management. Elements in this work queue are for example which contracts ends, which parts have multiple sources, which shipment accuracy is performed etc. Especially in Service Logistics the number of different suppliers has grown. Currently many different sources of supply exist which makes this process in particular difficult to manage.

9.2.5 Policy Fulfillment

The Policy Fulfillment process has three major objectives. The first objective is to support the parts availability. Parts availability is supported through defining and setting required stock levels, which lead to part movements in the network and part order requisitions. Also availability is monitored by early detection of part constrained situations and by executing a Fair Share Management process during part constrained situations. Furthermore availability is supported by the Policy Fulfillment process by interlocking with the Procurement organization and Source Management and Engineering organization to avoid or solve constrained parts problems.

The second objective is to support the Policy Planning process. This is done via the execution and management of the inventory plan in the most cost efficient way.

The third objective is to optimize the part cost. This is done via management of the planning supply hierarchy in the most cost efficient way, by applying the Order Mix Optimization process and by interlocking with the Service Parts Procurement and the Source Management and Engineering processes.

The service logistics networks are an important input for the forecast and planning processes. A distinction is made for the forecast and planning process at the central location and for the underlying part of the network. The forecast and planning process for the central location has focus on the flow of parts towards the network (hub) and on the planning of replenishments of lower echelons. In the flow of parts towards the network the order mix plays an important role.

Order Mix Optimization

To be able to fulfill the internal replenishment requests, the most economical options are reviewed by the part analyzers. Based on criteria such as type of source, part lead time, part price and availability, proposals are created for fulfillment orders. The type of source responsible or able to provide the part is an important decision making element. For example the part can still be in a warranty period and claimed from a supplier. It may also be possible to use excessive inventory from other geographies in the world wide network or is can be cheap the repair parts locally. Analyzers continuously match the part need and constraints and decide on the best type of order and quantity of parts.

It is checked if part costs can be claimed on the manufacturer if the parts are still within their warranty period. If the warranty period is exceeded, analyzers try to maximize the reutilization possibilities and propose orders on repair vendors. When no other alternatives are possible anymore a new buy order is proposed. The process of order type and part quantity selection that the analyzers perform is called the Order Mix Optimization. Once order proposals (also called requisitions) are created the actual order process starts.

Part Replenishments

The Part Replenishment process maintains the established stock levels in the network (from the hub downwards). In the past this was done from the hub to the Country Stock Room (CSR) and then further within a country from the CSR to the lower echelons. Improvements in this process have been made and in many cases parts are replenished directly from the central warehouse to the Regional Offices, Branch Offices and Support Stations[10]. Of course this saves handling, picking and packing costs in the supply chain.

Another important change in the Part Replenishment process had to do with integrating the stock of parts available in the various countries into the calculation of the amount of parts to order on the manufacturers and suppliers. In the past, planning for the central warehouse was based on a short term forecast ignoring the amount of stock available in the countries. An important study in this area (the World Wide Synchronization project) has been performed by IBM Amsterdam and IBM Mechanicsburg. The outcome of this study resulted in the architecture to synchronize the hub planning with the planning for the lower echelons of the network. Important elements in the solution were the creation of a long term aggregated forecast, a planning mechanism that includes the available stock in the network and an ordering mechanism that anticipates on the aggregated stock plan.

---

[10] This process is called 'Direct Branch Office Replenishment'.

9.2.6 Order Management

The Order Management process is also executed by the IBM Procurement organization. The process deals with the order placement, accounts payable, order monitoring and management processes.

Based on the order requisitions generated by the Policy Fulfillment process, orders are placed on various sources of supply as provided and maintained in the Source Selection and Contract Management process. Orders are received in the central location of the Service Network and moved internally in the Service Network to stock locations as appropriate.

In the Order Management process, criteria are defined and measurements have been created that measure the source shipment accuracy (i.e. the accuracy of the supplier lead time). The lead time (and variance in the lead time) is a leading parameter used in the planning process.

9.2.7 Source Management and Engineering

In the Source Management and Engineering process the technical relationships between the Service Logistics organization and the repair vendors are defined and maintained. Also decisions about the reverse logistics flows of defective parts are taken and maintained.

Because of defining and maintaining (technical content) relations with repair vendors in the Source Management and Engineering process also the parts quality is monitored and evaluated. If necessary, actions are started to solve quality problems.

Another important responsibility within this process is the monitoring of the cost savings as a result of the reutilization processes. This means that reutilization processes are enabled and disabled by Source Management and Engineering at the right moment in time. Providing different repair alternatives, makes it possible for the Order Management process to place orders on alternative repair vendors, generating saving on part costs.

9.2.8 Service Delivery

In the Service Delivery process the required parts delivery is handled. The process is triggered when a client contacts IBM via the Retain-linked Call Management System (RCMS). When a client contacts IBM with a service request, the request is evaluated and an entitlement check takes place. The entitlement check is meant to ensure that the type of service to be provided is aligned with the client maintenance contract. After the proper service type is identified, service part request information is interfaced to the Service Logistics process.

At the same time the service parts information is sent to the Service Logistics process a Client Engineer action is planned such that the Client Engineer and service part will both be available at the client site for maintenance purposes. The entitlement, call handling and the scheduling of the Client Engineer is currently not in the scope of the Service Logistics Management process. These matters belong more to the Service Delivery process.

When the service part request information is received from the call management system, actions within the Service Logistics process are triggered. In the part request a specific stock location is defined, which usually becomes the accepting stock location. If the part is in stock it will be allocated and delivered. In case it is not in stock an alternative stock location is searched. The search is done via so-called 'search paths'. A search path is a sequence of stock locations to be searched for the part availability. Different search paths are defined depending on the urgency type of the part request. Search paths start somewhere in the service network, following higher order stock locations until they end at the central location. The stock location that is actually delivering the part is called the executing stock location.

Depending on the type of part request (indicated as emergency type) and the transport method, appropriate actions are taken to ship the part to the required client (or other) location.

In case a requested part is not found in the applicable search path, a manual exception handling process is in place. This process is executed by the Central Emergency Desk (CED) and ensures that all alternatives available in the whole network are identified. Ad-hoc information is collected and via the CED an emergency order can be placed directly on a source of supply, from which the part request will be fulfilled.

9.2.9 Management Information System

The Management Information System (MIS) contains a wide range of tools and applications. The main objective of these tools is to provide management information on the performance of the organization. Measurements and Key Performance Indicators (KPI's) are important in the Service Logistics Organization. The reason for this is that contracts with clients usually contain performance indicators. The MIS applications are used to validate if the performance criteria are met and if the financial figures are at the correct level.

There are many measurements available in a Service Logistics organization. A few examples of these measurements are:

- Percentage of order requests delivered direct from shelf.
- Percentage of order requests delivered in the correct time frame.
- Inventory value.
- Operating cost.
- Number of replenishment orders outstanding.
- Transport cost.
- Parts cost (unit cost).

Table 9.1. Example of Management Information System measurements

Level	Measurement
Client	Client group – Client
Product	Product Type – Machine – Part
Location	Geography – Area - Country – Warehouse location
Time	Year – Quarter – Month – Week – Day

The measurements can be calculated at different levels. The level needed is dependent upon which purpose the measurements are used. Some examples of this are shown in Table 9.1.

In the last few years Business Intelligence (BI) solutions were developed and implemented as well. These tools enhance the flexibility of the reports and give management more insight than the predetermined reports. Because of BI solutions more dynamic and individualized reports can be created. Drill down capabilities were implemented which can be used for root cause analysis.

Due to historical reasons a variety of applications and measurements is used. Developments started within IBM several years ago show a tendency towards a global system of measurements and reporting tools. The globalization has the advantage that the key performance indicators are becoming identical. This makes it possible to compare the different geographies and create global reports and measurements.

9.3   System Solution Developments

The previous sections described the processes within the Service Logistics supply chain. Due to the globalization of the Service Logistics business, also systems are changed and developed to be part of a global system. In this section we give a brief overview of the most important elements of the Global Part System (GPS). Activities for developing GPS started several years ago.

GPS is a system consisting of various components. These components are developed by the IBM Amsterdam and IBM Mechanicsburg development centers. Besides own development also packages available on the market are integrated. Where support within GPS is not covered yet the legacy system environment is being used.

Systems used to support the Service Logistics supply chain processes are:

CPPS – Common Parts Processes and Systems

This system is the legacy system developed and used in EMEA only. The system supports many processes. All warehouses in EMEA are controlled by it except the central location, which is controlled by GPS.

A similar system, the Parts Inventory Management System (PIMS) is developed and used in the US and partially in other geographies.

The strategy is to integrate CPPS and PIMS in the GPS environment as much as possible.

GPS – Global Parts System

The GPS system is an integrated system environment including the Xelus-Plan and SAP packages (in particular the Material Management and Financials and Control modules). Currently GPS supports the logistics hubs (connecting the geographical networks). Developments are started to embed Network Neighborhood as the field planning application.

NN – Network Neighborhood

Network Neighborhood is a new IBM patented technology which supports the Demand Planning (or Stock Planning) process. The core of NN is a mathematical optimization model developed in cooperation with IBM and USA universities. The model is already implemented in the USA where specific sets of machines are controlled by Network Neighborhood. NN also calculates stock levels based on time based service requirements (Part Delivery Time). Preparations for implementation Network Neighborhood in other geographies are ongoing. It is to be expected that within the coming years Network Neighborhood will be the global primary supporting system for field stocking levels (with focus on the Same Day service network).

PSC – Policy Stock Calculator

This system was developed globally and is integrated into GPS. The system was developed to support the Early-Life forecast and planning. See also Draper and Suanet (2002).

MIS – Management Information Systems

The Management Information System is primarily based on a copy of the DB/2 production tables in a copy management environment. Furthermore data of several tables is combined in new tables (data warehouses), which are defined for specific purposes. This environment is referred to as the 'copy management' environment. On-line (electronic) table books are available in which the tables and data elements are grouped and described. A cross reference is included for efficient search procedures. Additional tools are installed with which data from the Copy Management environment can be used for reporting and analysis. Some examples of these tools are: DB/2 QMF, DB/2 OLAP Server, Executive Viewer and Brio.

9.4   The Next Step: Managing the Service Logistics Supply Chain

When we look at the Service Logistics Supply Chain, we can see dramatic changes the last few years. From recent developments several observation were made.

Firstly, differentiated on-demand logistics services are required. This is because clients are nowadays directly interacting with IT. As a consequence of this they have an increasing need for mission critical services. However the traditional process was more product-driven and considered specific client requirements to be exceptions. Also for base service the traditional service delivery model was too costly. Because of technology developments in products, e-service and e-service logistics, differentiated service solutions are enabled. It is for this reason that IBM Service Logistics has implemented and continues in developing and supporting an extensive service-palette. Examples of typical types of service provided are: Same Day Services (like 2, 4 and 8 hour committed repair services), Next Day services (usually overnight), On-Site stock, Client Replaceable Unit (CRU), Smart Couriers exchange, Depot repair (48 hours), Vendor Managed Ship (24 hours), Direct Client Ship (24 hours), Air Courier Ship (critical and very expensive parts) and Product Swap (24 hours).

Secondly, nowadays many market parties provide distribution networks, stock facilities and related services. Such a party is referred to as a Logistics Service Provider (LSP). Due to economies of scale these specialized companies can provide these services against lower costs.

Thirdly, many market parties offer dedicated part re-utilization processes. A continuous maintenance of the supplier information network is worthwhile and reduces the overall part costs.

Due to these developments IBM Service Logistics is more and more driven into a new role as the Supply Chain Manager, providing an on-demand and differentiated service in the most cost efficient manner while reducing the parts inventory. Global developments of processes, systems and changes in IBM's Service Logistics supply chain especially contributed in this direction.

Expected changes and further developments within IBM will have focus on the role as Supply Chain Manager. Specific developments can expected in the further implementation of the IBM patented Network Neighborhood model. This new technology is uniquely supporting the differentiated service approach for the high-end and mid-range market segments. In contrast to the traditional hierarchical network approach, Network Neighborhood utilizes the full service network structure. No predefined search paths are used during the forecast, planning and delivery processes. Instead the part requirements are matched with the whole network and an integral stock decision is made by the underlying mathematical decision model. While in a hierarchical approach a series of predefined stock locations will deliver the part, in Network Neighborhood all stock locations are considered and the best suited stock locations are determined dynamically and are recommended by the model.

Network Neighborhood combines the Client service requirements expressed in terms of Part Delivery Time, the service part needs as derived from the machine install base and the geographical positions (postal code areas) where the parts are required. Based on this information integral stock decisions are made by Network Neighborhood. As a consequence stock locations will consolidate inventory while satisfying the appropriate service constraints. As a result of the underlying mathematical decision model better stock decisions are made while service levels are maintained and costs are decreased.

References

Cachon, G.P. (2004) The allocation of inventory risk in a supply chain: Push, pull, and advance-purchase discount contracts, *Management Science* 50 (2): 222–238

Dekker, R., Fleischmann, M., Inderfurth, K., Van Wassenhove, L.N. (eds.) (2003) *Reverse Logistics: Quantitative Models for Closed-Loop Supply Chains*, Springer-Verlag, Berlin

Draper, M.W.F.M., Suanet, A.E.D. (2002) Service Inventory Optimization Algorithm, IBM Internal Paper, Amsterdam

Drezner, Z., Hamacher, H.W. (eds.) (2004) *Facility Location – Applications and Theory*, 2nd edition, Springer, Berlin

Fleischmann, M., van Nunen, J.A.E.E. and Gräve, B. (2003) Integrating Closed-loop Supply Chains and Spare-Parts Management at IBM, *Interfaces* 33(6): 44–56

IBM Global Service Logistics (1998) Worldwide Synchronization Study, Internal Study

Sherbrooke, C.C. (1992) *Optimal Inventory Modeling of Systems*, Wiley, New York

Silver, E.A., Pyke, D.F., and Peterson, R. (1999) *Inventory Management and Production Planning and Scheduling*, 3rd edition, Wiley, New York

Simchi-Levi, D., Wu, S.D., Shen, Z.M. (eds.) (2004) *Handbook of Supply Chain Analysis in the E-Business Era*, Kluwer Academic Publishers

Teunter, R.H., Klein Haneveld, W.K. (1998) The 'final order' problem, *European Journal of Operational Research* 107(1): 35–44

10 Business Process Integration

Santhosh Kumaran and Kumar Bhaskaran

10.1 Introduction

Supply Chains are highly competitive and ultra responsive business models that integrate information and decision across all participants in an extended enterprise. In today's global economy, enterprises are changing continually, entering into new markets, encountering new competitors, introducing new products and restructuring themselves through mergers, acquisitions, alliances and divestitures. In order to stay competitive in such environments, enterprises require supply chain management solutions that are agile, responsive, resilient, and dynamic. Business process integration and management (BPIM) constitutes a set of technologies that serve as the foundation for creating such solutions. This paper presents a vision for the future supply chain systems, identifies the technical challenges in realizing this vision, and outlines a solution leveraging BPIM technologies.

10.1.1 The Vision

We envision the future supply chains to be virtual enterprises "whose business processes—integrated end-to-end across the company and with key partners, suppliers and customers—can respond with flexibility and speed to any customer demand, market opportunity or external threat" (Palmisano 2002). Disjointed processes, unaligned applications, and traditional communication channels, however, still burden a number of supply chains today. Bridging this gap will require supply chain solutions that support integrated processes, enable inter-operable applications, and provide unfettered access to heterogeneous data sources to present an end-to-end solution image. Any new supply chain solution must also carry forward the significant business investments in legacy applications and data sources, be flexible to accommodate changes in business processes through new integrated business functions, and enable collaboration across enterprise boundaries.

Business process integration and management (BPIM) has emerged as a new paradigm for architecting, developing, deploying, and managing the next generation business operation systems for supply chains that flexibly and efficiently combine content, process, and people within and between enterprises to effectively manage the supply chain as a virtual enterprise. The emergence of BPIM to realize supply chain solutions is the result of the convergence of a number of business and technology factors.

The global competitive battles today are between business models as manifested in the overall Supply Chain (SC). This requires businesses to be able to forge trading partner relationships and integrate their respective business processes to respond to the market. Businesses would very much like to focus on this prime directive rather than be concerned about myriad applications, data sources, and middleware that make up the SC information system. BPIM through its model driven integration approach enables business users to focus on the core capabilities of the business rather than on the complexities of the underlying integration technology. From a technology standpoint, the maturity of distributed object computing technology and the rapid rise and acceptance of Internet technologies makes it possible to capture in software the SC vision using BPIM.

10.1.2 The Challenge

Supply chains today are dysfunctional in many respects. Business processes continue to be reengineered to ensure that they are streamlined and integrated. A number of disparate applications are employed to support the business processes. Often these applications are not inter-operable. The joint effect of disconnected processes and applications is the well known "functional silos" that characterize the supply chain. Additionally, the supply chain community uses traditional communication channels such as fax, telephone, and e-mail to collaborate.

Businesses recognize the limitations of the current SC environment and its deleterious effects on competitiveness and overall business performance. There is therefore a clear business need for SC solutions that bring enterprises closer to realizing the SC vision described previously.

10.1.3 Bridging the Gap

Contrasting the SC reality with the vision highlights a number of fundamental desirable capabilities for SC solutions:

- Integrated Solution: Businesses cannot function efficiently with disjointed processes and applications. SC solutions must support integrated processes, enable inter-operable applications, and provide unfettered access to heterogeneous data sources to present an end-to-end solution image.
- Leverage Existing Assets: Businesses have significant investment in legacy applications and data sources that have to be carried forward in any new SC solution. This implies that the SC solutions are likely to be composed rather than built completely new from scratch.
- Cross-Enterprise Enabled: Supply Chains involve multiple trading partners. Consequently SC solutions must be capable of automating business processes that cross enterprises (e.g. collaborative forecasting and replenishment planning, collaborative engineering and product design), and enable collaboration among role players (human as well as applications) across enterprises.
- Flexibility to Adapt: SC is a dynamic business system. As business processes change, which they inevitably do, the SC solution must be flexible to accommodate the change. Such adaptation can be in the form of new integrated business function to support the process change. Monolithic applications that are hard wired are obviously inflexible. Consequently, it is typical to observe applications in today's SC that are unaligned with the business processes.
- Electronic Commerce Ready: Historically commerce has been closely tied to advances in communication. The Internet represents a significant convergence of communication and computing that permits exchange of information in real time, anytime, and anywhere in a global scale. This new way of doing business, Electronic Commerce, is a SC solution enabler as it permits the real time integration of SC across enterprises. SC solutions must be capable of leveraging the advances in Electronic Commerce to be viable to businesses in the future.
- Bridge the business-IT gap: As the use of IT has become ubiquitous in the enterprise and its capabilities has expanded enormously, there is ever increasing demand for its ability to directly support the business processes of the enterprise. Supply chain management solutions should directly link the business goals and objectives of the enterprise with the IT solutions.

These requirements suggest a flexible assembly model for providing an integrated SC solution. Business Process Integration and Management (BPIM) technologies provide just such a model.

10.2 A Multi-Layer Modeling Framework for BPIM

Explicit modeling of the structure and behavior of business systems is at the core of BPIM technologies (Cyert and March 1992; Steven 2003). We present an innovative, multi-layer modeling framework for BPIM that serves as a powerful vehicle for creating agile, responsive, resilient, and dynamic supply chain management solutions (Kumaran 2004). The framework is made up of four layers: Strategy, Operations, Execution, and Implementation (Figure 10.1). Each layer constitutes a different level of abstraction, performs a well-defined function, and has a different audience. Strategy layer defines the goals and objectives of the supply chain from the perspectives of the key stakeholders. Operation layer describes the business operations performed by the supply chain participants to achieve the goals. Execution layer is an abstraction of the computational elements that are needed to execute the business operations. Implementation layer specifies how the computational elements are implemented on a specific IT platform. Below we briefly introduce each layer and discuss the connections between the layers.

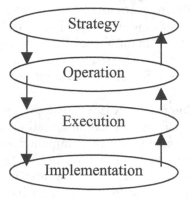

Fig. 10.1. Framework Core

Strategy Layer: The models built at this layer are used to specify what the business wants to achieve. It models the business objectives in terms business leaders understand. For example, it might specify the objectives in terms of the well-known Balanced Scorecard perspectives (Kaplan and Norton 1992):

- Financial Perspective - How should we look to our shareholders?
- Internal Business Perspective - What must we excel at?
- Innovation & Learning Perspective - How can we continue to improve & create value?
- Customer Perspective - How do we want our customers to see us?

Thus, the model is initially expressed in terms normally used in business strategy discussions, and is verified through iterative interaction with strategy executives.

Operation Layer: The Operations Model describes what the business is doing to achieve its strategic objectives, and how it will measure its progress toward them. It is typically developed by operations executives in collaboration with strategy executives. It captures the business operations, commitments, and Key Performance Indicators (KPIs). The KPIs are directly linked to the measures that indicate progress on Balanced Scorecard goals. A few examples are given below:

- Financial – What is our margin per SKU?
- Internal Business – What is the average number of days to process a new business opportunity?
- Learning & Growth - What suppliers have an average response time that is degrading?
- Customer - How fast are we responding to customer change requests?

The KPIs are linked to specific business operations and processes. The KPI are expressed in terms that operating executives understand.

Execution Layer: The platform-independent Execution Model describes the behavior and structure of the computational models used to implement the business operations. It does not assume a particular implementation, allowing iterative performance improvement while assuring consistency with the business objectives. A transformation tool is used to create the core elements of the Execution Model from the Business Operations Model and then manually refined to create a complete definition of the execution model.

Implementation Layer: The platform-specific Implementation Model defines the actual IT processes in a specific realization of the Execution Model. Tools are used to construct portions of the Implementation model directly from the Execution Model much as a compiler translates a high-level language. The model links to applications and specifies how to measure the parameters needed to determine the KPIs. Mapping from the implementation model to an actual implementation may happen either via code generation or through scripting.

10.3 Supply Chain Scenario

This section describes a customer scenario to serve as a running example throughout the document for illustrating the detail workings of the BPIM framework. The scenario deals with Just-In Time Scheduling.

An electronics manufacturer has a committed production schedule for the next several (3) days, but a customer has made a request to increase his order significantly, and that order is due to be produced within the next 3 days. Referring to the Solution Context Diagram (Figure 10.2), the customer inquiry has been input to the manufacturer's fulfillment system, which initiates an Available-to-Promise Check by submitting a Build Plan Change Request to the Just-In Time Scheduling process. The Fulfillment System has already checked the build plan and determined that the order change cannot be accommodated by the current build plan. (The build plan did not call for producing sufficient quantity of the ordered product for the customer or to stock to meet the increased demand.) The manufacturer needs to check whether he has the capacity and materials to meet the request. If the supply plan (which includes inventory) does not have enough of each required part, he needs to contact suppliers to find out whether they can provide the additional parts needed to manufacture the increased quantity of product. First he requests the parts from his primary suppliers; only if they cannot meet the request does the manufacturer look for parts on the spot market. If all the parts can be acquired, he decides whether or not to accept the increased order (in part based on the quoted prices of the parts). If he decides to take the order, he confirms with his suppliers, updates the build plan, and responds to the customer. The manufacturer uses the results of the supplier interactions to update the suppliers' profiles for future reference.

The Just-In Time Scheduling process is important to the manufacturer for the following reasons:

- Satisfy customers by accommodating their change requests
- Improve production capacity utilization
- Quickly identify alternative sources of supply
- Dynamically update supplier evaluations

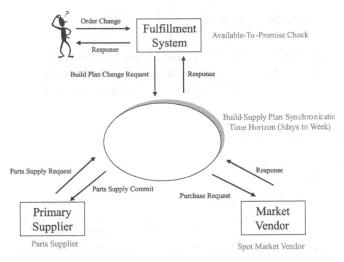

Fig. 10.2. Context Diagram for the Customer Scenario

Serviceability is an aggregation of various process metrics such as availability, reliability, and response time. Fill rate is the percent of order changes that can be accommodated. Supplier flexibility is a measure of how frequently the primary suppliers can meet the increased demand. Supplier performance is a measure of how well the suppliers actually perform, both in committing to the change and delivering as promised.

Applications that the manufacturer uses in this process are:

- Manufacturing Resource Planner (MRP)
- Manufacturing Scheduling (Build Plan)
- Supply Scheduling (Supply Plan)

The solution should manage the entire Just-In Time Scheduling process. The system should automate certain steps in the process, and coordinate the manual workflow steps. The system should be integrated with the fulfillment system, which initiates the process with a Build Plan Change Request and requires a prompt response. The applications that check and update the build plan, the supply plan, and the vendor profiles must be integrated into the system. The system should automatically communicate parts requests and responses with the primary suppliers using message protocols and formats defined in the suppliers' profiles.

During the business transformation (BT) process itself, it is important that the solution is "plumbed" for the capabilities enabled by the framework; e.g., monitoring, management, adaptability, etc. Furthermore, the data collected while the solution is "live" is critical input to the next BT.

10.4 Framework Details

In this section we describe the details of the multi-layer BPIM modeling framework. We present the key modeling elements in each layer of the framework and give examples from the JIT scheduling process.

10.4.1 Strategy Layer

The strategy model defines the "goals and objectives" of the business entity. We focus primarily on the management aspect of the strategy. We leverage the balanced scorecard work (Kaplan and Norton 1992) in defining the strategy management model. The key modeling element is called Scorecard. The four balanced scorecard perspectives (Financial, Customer, Internal Processes and Learning & Growth) can be derived from it.

Each perspective is a collection of 4-tuples. Each 4-tuple consists of the following elements:

1. "Objective" defines the business objective managed by this element of the perspective.
2. "Measure" defines the strategic KPIs that are used to track the progress on this objective.
3. "Target" defines a vector that shows expected values and milestones of the measure
4. "Initiative" defines the business operations in place to support the objective.

The objectives may have dependencies among themselves. These are captured using a set of causal links between the tuples. The model allows the definition of constraints on the measures as well.

Figure 10.3 shows the balanced scorecard for the manufacturer.

10.4.2 Operation Layer

Operation models present the perspective of the Line-Of-Business (LOB) managers. These models constitute the abstractions needed to describe the business operations that the business employs to achieve the strategic goals. These models include the monitoring and management mechanisms to ascertain the effectiveness of the operations in achieving these goals. The metamodel for the operation layer is defined based on the following observations:

- Operations are different from automations. The operation models should facilitate an execution strategy in which only some of the operations are automated.
- Operations are different from business process executions. Thus even those operations that are automated may be executed in multiple ways. The operation model should not include execution semantics.
- The stakeholders of the operation models are LOB managers. The models should project a view of the business operations at a level of abstraction that is suitable for this audience. This implies the importance of the "right granularity".
- Operation models provide a description of the business that the business owner can use to manage and control the business over its entire life cycle.

The models in the operation layer are organized as consisting of a base model with decorator models attached to it using the Decorator design pattern (Gamma et al. 1995). There are three decorator models: Governance model, Simulation model, and Organization model. The Operation Layer design is influenced by the work on Operational Specifications described in Nigam and Caswell (2003).

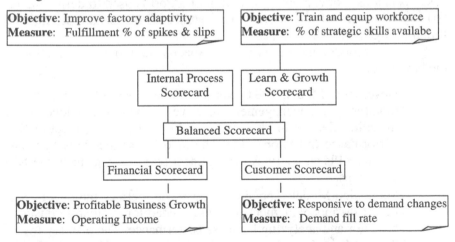

Fig.10.3. Balanced Scorecard for JIT Example

Business Operations are modeled as a factorization of the operational knowledge into two pieces:

1. Business Task: A Business Task is an "atomic" piece of the operation as seen by a business person. A Business Task cannot be decomposed any further at the business level.

2. Artifact Repository: An Artifact Repository is a store or a "wait shelf" where Business Artifacts can be placed as part of the operation. Artifacts stay in an Artifact Repository till a Business Task explicitly extracts them. Business Artifacts contain information pertinent to the business operation, as one concrete collection of information, and thereby form the domain of the business operations.

Business Tasks are "in the business of processing" Business Artifacts. Business Repositories, on the other hand, are "in the business of storing" Business Artifacts. Business Artifacts are the fundamental factorization that a business person uses to talk about all aspects of the business e.g. "This is what I am in the business of producing or processing or storing".

A Business Operation is an aggregation of Business Tasks, Artifact Repositories, and possibly other Business Operations. A Business Operation is represented as a connected graph with the business tasks, artifact repositories, and other business operations as nodes and the flow of artifacts between these elements as edges. A business operation can be contained in another business operation and can also overlap with another business operation. Two business operations that overlap contain some common business Tasks and/or artifact repositories.

The Metamodel has elements that allow for precise definition of business operations. For example, the notion of a Port is used to define the interfaces of tasks and repositories. The details of the metamodel may be found in Nigam and Caswell (2003).

A set of decorator models is used to specify additional information to the base model.

- Governance Model: The Governance model is used to describe the monitoring and management directives at the business level. Core modeling element in the Governance model is the concept of Key Performance Indicator (KPI). The model describes how to measure the KPIs and defines management policies to manage the KPI deviations.
- Simulation Model: The simulation model defines the characteristics of the model elements that are needed for simulating business behavior and analyzing business performance. The primary use of the simulation model is to simulate business operations.
- Organization Model: The Organization Model defines the organizational role players that participate in the business operations.

10.4.3 Operation Model for the JIT Example

Figure 10.4 shows the operation model for JIT scheduling.

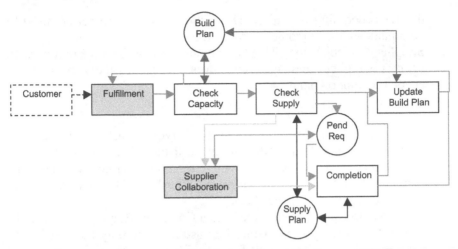

Fig. 10.4. Operation Model for JIT Scheduling

The JIT scheduling process is represented as a connected graph consisting of three artifact repositories, four business tasks, two internal business operations, and an external operation at a customer site. The fundamental abstractions are the business artifacts that are processed by the tasks and stored in the repositories. There are four business artifacts as listed below:

1. Customer Request: This is the key business artifact of the JIT scheduling process. The goal of processing this artifact is to ensure the availability of manufacturing capacity as well as the parts needed to meet the customer request. Once the results of the checks, positive or negative, have been recorded on the artifact, the processing is complete. The edges that represent flow of this artifact are colored red.
2. Parts Request: This artifact lists parts needed, but not on hand, to satisfy the customer request. The edges that represent flow of this artifact are colored brown.
3. Build Plan: This records the allocation of manufacturing capacity. The edges that represent flow of this artifact are colored blue.
4. Supply Plan: This records allocation of parts inventory. The edges that represent flow of this artifact are colored black.

The two internal business operations are as follows:

1. Fulfillment: This operation creates a Customer Request artifact in consultation with the customer. These interactions with the customer are represented as non-artifact exchanges of information. Once a Customer Request has been created, it is placed on the output port. At a later time, this operation receives a fully processed Customer Request from

the JIT scheduling operation. The contents of this artifact are used to communicate the outcome to the customer.

2. Supplier Collaboration: This operation handles the negotiation with primary suppliers and market vendors to procure the parts that are needed but not on hand.

The four Business Tasks are as follows:

1. Check Capacity is triggered when it receives a Customer Request. It consults the Build Plan artifact to check if there is available manufacturing capacity. The decision is recorded on the Customer Request artifact, and the artifact is sent out on the "Yes" or "No" port depending on the determination of the check.
2. Check Supply starts when it receives a Customer Request with an indication that capacity is available. It consults the Supply Plan artifact to determine if the parts needed for manufacturing the widget are available. If parts are available, this fact is recorded on the artifact and it is sent out to the next task, Update Build Plan. Otherwise, a Parts Request artifact is created and sent out to the Supplier Collaboration operation. Additionally the original Customer Request artifact is updated to reflect the shortfall and stored in the "Pending Requests" repository.
3. Completion task starts when the "Supplier Collaboration" operation updates the Customer Request artifact in the "Pending Request" repository with the final availability information. It accesses the corresponding processed Parts Request artifacts from a repository maintained by the Supplier Collaboration and makes a final decision on whether to fulfill the customer request. If the decision is "Yes", the request is sent to the "Update Build Plan" task. Otherwise, the request is returned to the Fulfillment operation. This task also modifies the Supply Plan as applicable.
4. Update Build Plan starts when it receives a Customer Request where the results of the supply check and parts availability are recorded as positive. It updates the Build Plan and marks the Customer Request as "done", and places it on the output port connected to the Fulfillment operation.

There are three Artifact Repositories:

1. Pending Request contains Customer Request artifacts with a positive result for the capacity check and a "maybe" determination of the supply check, i.e. these are requests for which parts are not available in the parts inventory.
2. Build Plan holds the current manufacturing schedule.
3. Supply Plan holds the current inventory and the parts that are already in the procurement pipeline.

Figure 10.5 shows the operation model for Supplier Collaboration in JIT scheduling process. It is made up of two business tasks, two external operations, and three artifact repositories. In addition to the four business artifacts introduced earlier, this operation includes Vendor Profile which records vendor information including terms and conditions.

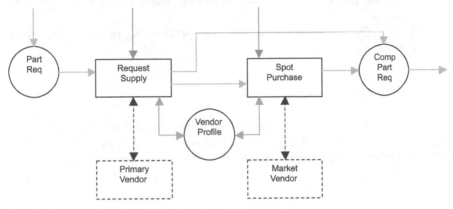

Fig. 10.5. Operation Model for Supplier Collaboration

The business tasks in the supplier collaboration are as follows:

1. Request Supply starts when a Parts Request is available in the Parts Request repository. It negotiates with the Primary Vendor, using the information in the Parts Request and the Vendor Profile, using non-artifact exchanges. There are two possible outcomes of this interaction: If the Primary Vendor agrees to provide the parts requested, the Customer Request in Pending Request repository is modified, the Parts Request is modified to reflect the vendor's agreement and is sent out to the "Completed Part Requests" repository. If the Primary Vendor cannot supply the parts, this fact is recorded on the Vendor Profile, and the modified Parts Request is sent out to the "Spot Purchase" task.

2. Spot Purchase starts when it receives a Parts Request that could not be filled by the Primary Vendor. This task now negotiates with a Market Vendor through non-artifact exchanges to acquire the parts needed. There are two possible outcomes, however, in both cases once the task is complete, the Parts Request artifact is sent out on the output port to the "Completed Part Requests" repository. If the Market Vendor agrees to provide the parts requested, the Customer Request in Pending Request repository is modified, the Parts Request is modified to reflect the vendor's agreement and is sent out on the output port. If the Primary Vendor cannot supply the parts, this fact is recorded on the Vendor Profile, and the modified Parts Request is sent out on the output port.

The artifact repositories are:

1. "Parts Requests" repository stores the requests for parts created by the Check Supply task in Fig 4.
2. "Completed Parts Requests" repository contains Parts Requests on which the primary vendor and/or market vendor decisions have been recorded.
3. "Vendor Profile" repository stores the profiles for primary and market vendors.

10.4.4 Execution Layer

There are three key abstractions in the Execution Layer: Adaptive Business Objects, Activity Flows, and Connectors. A Solution Composition model enables composition of end-to-end solutions from these components. Security and solution management models are defined as decorations over the base model.

Adaptive Business Object

An Adaptive Business Object (ABO) is the execution level abstraction of the business artifact introduced in the previous section (Nandi et al. 2003). It models the structure and behavior of the artifact. The key elements of the ABO metamodel are given below (Figure 10.6):

- Lifecycle & Behavior: The lifecycle of the business artifact is defined using a finite state machine (FSM). The states of the FSM correspond to the lifecycle states of the artifact. An ABO receives external events via its public interface and reacts to these events via action invocations triggered as part of the state transitions.
- Data Graph: Unlike traditional business objects, data is not contained inside an ABO. Instead, ABO uses a data graph to dynamically aggregate information on demand from heterogeneous data sources. Aggregation and presentation of data in this manner is just another manifestation of the ABO behavior and thus influenced by the state of the ABO. The graph structure implicitly enforces the data relationships and their cardinalities. This abstraction provides the modeler with the ability to specify a data model irrespective of the physical store.
- State Adaptive Access: People and applications may interact with the ABO at various points in its life cycle. As part of such interactions, parts of the data graph may be accessed or manipulated or business events may be sent to the ABO by invocations of its pub-

lic interface. However, the ability of the outside world to raise events or access the ABO data is determined based on the current state of the ABO. The representation of the entity lifecycle by state machine provides the ability to specify state adaptive access control to business roles. The modeler can specify read, write, search access on data and authorize access for events for each business role and state combination.

– Data Actions: The Data Actions are used to model CRUD operations on parts of or the whole data graph. These are invoked by the ABO as a side effect of state transitions. This effectively enforces the constraints regarding data integrity, transaction scope, access control, and business semantics.

– Remote Actions: An ABO changes its environment via Remote Actions fired as part of state transitions. These are defined using the Command design pattern (Gamma et al. 1995). The model can generate the appropriate Web Services Definition Language (WSDL) definitions (Christensen et al. 2001) for the command interfaces and bound the receivers to any network accessible service during deployment. In the business process context, such services could include workflow processes and enterprise information systems.

– Views: Views present the external interface or API of the ABO. There are 3 main components that constitute the View.

1. Query. This enables the user to search for ABO instances satisfying certain criteria. Searchable fields include the current state and the data graph. The query returns a list of ABO instances.
2. Data. This is the interface to obtain partial or the whole data graph of a particular ABO instance.
3. Events. These specify the events accepted by the ABO and the corresponding event parameters.

For human users, the screens to drive user navigation can be automatically generated based on the state adaptive access. Thus there will be a 'view' for each ABO state for each interacting business role.

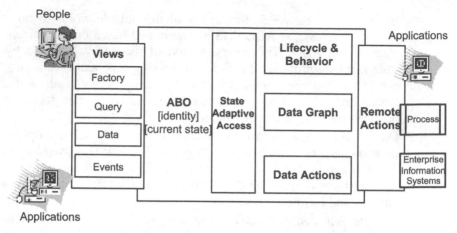

Fig. 10.6. Elements of the ABO metamodel

Activity Flows

Activity Flow models the structure and behavior of an activity leveraging the activity behavior model in UML2 (OMG 2003). The execution semantics is based on Workflow Petri Nets (Leymann and Roller 2000). At the core of the activity flow model is the ability to define a control flow as a directed acyclic graph (DAG). The model elements are as follows:

–   Operation: An operation invocation that is not further decomposed in the model. Operations are attached to the nodes of the DAG.
–   Activity Flow: A node in the DAG could be an Activity Flow. This enables hierarchical composition of activity flows.
–   Sequence: Control edges that define the sequencing of steps in the activity.
–   Control nodes: Execution steps that are purely meant for control information. This includes Start, Stop, Fork, Join, Decision, and Merge.

Activity Flows are the execution level abstractions of business tasks in the operation layer.

Connectors

Connector models are used to define the structure and behavior of the components that integrate human users, business partner systems, enterprise information systems, and applications with business processes. The goal is to use a single metamodel to define the core behavioral aspects of a connector that covers all of the above integration scenarios. The definition of the connector metamodel is influenced by the work on conversational support for component integration (Hanson et al. 2002).

The B2B connectors are used to link business processes across enterprise boundaries. EAI connectors are used to integrate enterprise applications with the business processes. HCI connectors model Human-Computer Interactions, enabling people to participate in the business processes. The behaviors of all the three connectors, namely, EAI Connector, B2B Connector, and HCI Connector, are described in terms of the following aspects:

- The roles played by the parties that are connected by the connector; the party could be a person, a program or an enterprise.
- The message sets exchanged between the parties. If the parties are using different message sets, the Connector needs to do the necessary transformations. Typically, a Connector defines a canonical message set and transforms the incoming messages to this format and from this format to the outgoing messages.
- The sequencing and timing constraints on the message exchange. This may be thought of as the protocol used by the involved parties to interact with each other.

Security

The execution level security model deals with the security aspects of the IT level solution and its components. It covers issues including quality of protection, authentication, authorization, data integrity, confidentiality, and non-repudiability. The security model is designed to permit top-down specification of security requirements. With this approach a security architect can define the requirements on abstract components, and subsequently map the requirements to deployment configurations that are applicable to the realizations of those components.

Solution Composition

Model-based Components are the building blocks of model-drivel solution creation. The components include Connectors, ABOs, and Activity Flows. The execution model describes the behavior of these components. We create a Solution Artifact from a component by annotating it with a service description. Thus Solution Artifacts are service providers that participate in a service-oriented architecture. Solution Templates may be composed from several Solution Artifacts. Solution Artifacts and Solution Templates are platform-independent. Platform-specific implementations may be provided for Solution Artifacts or for an entire Solution Template (Huang et al. 2004).

The model elements below are used to compose end-to-end solutions from the components discussed in earlier sections.

- Participants: List the participating solution components and their Implementations
- Direct Link: The direct invocation relationship between two (Source and Target) components.
- Event Link: The event pub-sub relation between components.

Execution Model for the JIT Scheduling

The components that make up the execution model for JIT scheduling process are as follows:

There are two ABOs:

1. Customer Request: This models the end-to-end lifecycle of the customer request that triggers an instance of the scheduling process.
2. Part Request: This models the end-to-end lifecycle of part requests generated by the manufacturer in response to a customer request when additional parts need to be procured from the vendor.

There are five Activity Flows:

1. Check Capacity: This models the activities and their sequencing for checking the production capacity in the context of fulfilling an order for some widget.
2. Check Supply: This models the activities and their sequencing for checking the inventory to determine if parts are available to produce the ordered widget.
3. Request Supply: This models the activities and their sequencing to collaborate with vendors towards procuring the parts needed to produce the widget.
4. Spot Purchase: This models the activities and their sequencing in the context of the spot purchase task.
5. Completion: This models the activities and their sequencing in the context of the completion task.

There are seven Connectors:

1. Market Vendor: This is a B2B connector that drives the business-to-business interactions between the manufacturer and a market vendor in the context of a spot purchase activity.

2. Primary Vendor: This is a B2B connector that drives the business-to-business interactions between the manufacturer and a primary vendor in the context of the procurement activity.
3. Fulfillment: This is an EAI connector that handles the integration of fulfillment application with the business process.
4. Supply Plan: This is an EAI connector that handles the integration of supply plan database with the business process.
5. Build Plan: This is an EAI connector that handles the integration of build plan database with the business process.
6. Scheduler: This is an HCI connector that enables the Scheduler role player to interact with the business process.
7. Purchaser: This is an HCI connector that enables the Purchaser role player to interact with the business process.

Figure 10.7 shows how the components are composed to create an end-to-end execution model for the JIT scheduling process. The ABOs serve as "service brokers", the Activity Flows serve as "service compositions", and the connectors serve as interfaces to services.

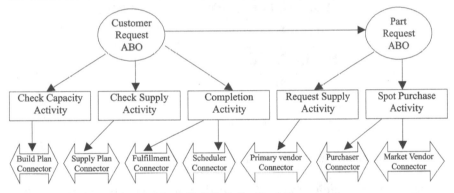

Fig. 10.7. Execution Model for JIT Scheduling

10.5 Sense and Respond Supply Chains through BPIM

The business process integration and management framework introduced in this paper provides the technology underpinnings for the "sense and respond" supply chains. At its core, a sense and respond business system is one that can detect significant business events or situations from the "enterprise event cloud" (Luckham and Frasca 1998) and respond to these events in a timely manner (Haeckel 1999) The multi-layer models by which we describe the structure and behavior of a supply chain play an important role in enabling a supply chain partner to detect significant business events and to respond to these events promptly.

As discussed in the earlier sections, the models capture the KPIs and their relationships at various levels. The strategy model defines the scorecard measures, the operation model defines the process-level KPIs, and the execution model defines the probe points that instrument the runtime components to collect the raw monitoring data. Leveraging the connections between the layers and the context information contained within the layers, we can aggregate and correlate low level raw events all the way to the KPIs at business operation level and strategy level (Figure 10.8).

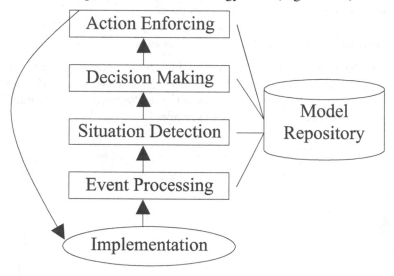

Fig. 10.8. Model Driven Supply Chain Performance Management

Management policies may be attached to any KPI. These policies specify the actions to be taken when certain business situations are detected. These actions are specified as business operations and the BPIM framework can be used to implement these operations. Thus the BPIM framework enables a "sense and respond" system by combining the top-down model-driven business integration with bottom-up model-driven business process management as shown in Figure 10.9.

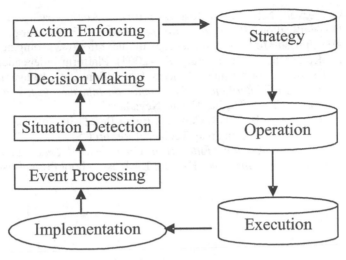

Fig. 10.9. Sense and Respond Supply Chains

References

Palmisano, S. (2002) The New Agenda, available: http://www.ibm.com/
Christensen, E., Curbera, F., Meredith, G., Weerawarana, S. (2001) Web Services Description Language [Online], available: http://www.w3.org/TR/wsdl.
Cyert, R.M., March, J.G. (1992) *A Behavioral Theory of the Firm*, Blackwell, Cambridge, Massachusetts, Second Edition
Kumaran, S. (2004) Model Driven Enterprise. *Proceedings of the Global EAI Summit*
Gamma, E., Helm, R., Johnson, R., Vlissides, J. (1995). *Design Patterns – Elements of Reusable Object Oriented Software*, Addison-Wesley Publishing Company, NY
Hanson, J.E., Nandi, P., Kumaran, S. (2002) Conversation Support for Business Process Integration, *Proceedings 6th ÍEEE International Enterprise Distributed Object Computing Conference (EDOC-2002)*, IEEE Press, 65–74
Kaplan, R.S., Norton, D.P. (1992) The Balanced Scorecard – Measures That Drive Performance, *Harvard Business Review*, Jan–Feb, pp 71–79.
Kimbrough, S. (2003) Computational Modeling: Opportunities for the Information and Management Sciences, *Eighth INFORMS Computing Society Conference*, Chandler, Arizona
Leymann, F., Roller, D. (2000) *Production Workflow: Concepts and Techniques*, Upper Saddle River, New Jersey: Prentice Hall PTR.
Nandi, P., Kumaran, S., Chung, J.-Y., Heath, T., Das, R., Bhaskaran, K. (2003) ADoc-Oriented Programming, *International Symposium on Applications and the Internet*, Orlando, Florida

Nigam, A., Caswell, N.S. (2003) Business artifacts: An approach to operational specification, *IBM Systems Journal*, Volume 42, Number 3, Page 428.

OMG (2003) Unified Modeling Language, available: http://www.omg.org/

Huang, Y., Kumaran, S., Bhaskaran, K. (2004) Platform-Independent Model Templates for Business Process Integration and Management Solutions, *IEEE International Conference on Information Reuse and Integration (IRI 2003)*, Las Vegas, Nevada

Luckham, D.C., Frasca, B. (1998) Complex Event Processing in Distributed Systems. Stanford University Technical Report CSL-TR-98-754

Haeckel, S.H. (1999) *Adaptive Enterprise: Creating and Leading Sense-And-Respond Organizations*, Harvard Business School Press, Boston, MA

11 Collaboration in e-Supply Networks

Chris Nøkkentved

11.1 Introduction

Markets once favored competitors that could successfully integrate massive horizontal or vertical asset bases to create economies of scale. The current global environment, marked by increased demand, decreased customer loyalty, shorter product life-cycles, and mass product customization, forces companies to lower costs while increasing the quality and variety of products and services. The rise of Business-to-Business (B2B) Trading Networks over the Internet enables companies to meet these challenges by extending their value-chains and cooperating with organizations whose complementary capabilities can give the whole business network a competitive edge. The ability to share, integrate and collaborate with other businesses provides an additional differentiation for companies competing with large asset-based competitors. The need to better integrate with customers and suppliers compels businesses to dramatically alter their processes in order to survive. As the cost and latency or friction is removed from B2B transactions, companies will be more willing to consider outsourcing what were once core business processes, thus finding themselves as participants in multi-company business processes. Consequently, many companies are currently disassembling their process infrastructures into independent processes and then reassembling them as parts of an extended supply network via outsourcing and collaborative partnerships, thus concentrating on their core competencies and process capabilities. This kind of partnering might also mean working collaboratively to share production, demand, capacity or product information in order to synchronize business behaviors across a supply network.

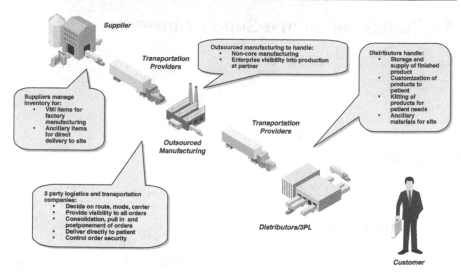

Fig. 11.1. Collaborative Planning Leverages of Partner Skills

Industrial competition is therefore advancing from being between individual companies, to being between clusters of tightly-knit partners with the intent of delivering to the customer the desired product within a fitting time-frame, at the right price. Hence, companies are progressing from the notion of the extended supply chain and supply networks into e-Supply networks facilitated by electronic B2B marketplaces (Trade Exchanges or e-Markets). These support inter-organizational information sharing, transactional integration and collaborative, event-driven processes taking place in bilateral and multilateral relationships between partnering firms. By conducting Collaborative Planning within such a company's own Private e-Market or via Public e-Markets (e.g. consortiums), companies attempt to operate their value-added communities as though they were one seamless organization, synchronized to meet customer demand, in their pursue to achieve significant cost savings and service enhancements.

This chapter will attempt to provide an overview of collaborative relationships and processes within e-Markets that utilize the Internet to facilitate co-ordination and enable collaboration among multiple trading partners. We will expose how e-Markets are currently deploying supply chain planning applications that bind firms through information-sharing, interdependent transactions and collaborative processes. Then we will delve into the various collaborative planning processes that may take place within an e-Market. Finally, we will present some of the benefits and implementation considerations of collaborative planning beyond the largely descriptive and anecdotal presentation of the advantages of e-Business from popular literature and the press.

11.2 From Supply Chains to e-Supply Networks

The dawn of the new digital, networked economy[11], enables enterprises to transform themselves into *adaptable processes networks*[12]. The advent of the Internet as a universal communications platform extends even further a company's reach, and enables richer information exchange among collaborative networks of partners. In such an environment, companies must be *flexible* and *agile*—able to react quickly with minimal effort and expense. Agility can be greatly increased by improving the ability to detect problems, threats, and opportunities, giving the organization and its partners more time to react. Innovative companies are using current advances in information technology[13] (like Collaborative SCM systems) and utilize

---

[11] Among the most noteworthy proponents of the "digital economy" are Tapscott (1995), and Shapiro and Varian (1998).

[12] The concept of adaptable process networks was presented by Chisholm (1998).

[13] The appearance of such ephemeral "plug-and-collaborate" supply networks and virtual B2B collaborative communities (e.g. E-Business/Trade exchanges), has been enabled by innovative advances in information technology and driven by the utilization of common communication, security and process standards. In the last decade a rising number of companies have been experimenting with process improvement, integration and automation. Most of these business engineering efforts were realized via enterprise resource planning - or ERP systems (from vendors like SAP, Baan, PeopleSoft, Oracle). The wider deployment of ERP systems and innovations in messaging and tracking technologies that allow real-time management of supply chain activities, has resulted in more compatible process and information infrastructures. Furthermore, the Internet has emerged as an ubiquitous communication platform on which companies can collaborate with their partners, reduce cycle times and enforce data and security protocols. These developments have led to the appearance of advanced planning, optimization and scheduling software (APS) that complements ERP/ MRP with an intelligent planning environment. APS/SCM systems implement supply network planning processes and act as a highly responsive nervous system of a supply network. Many software vendors are currently offering SCM systems (e.g. SAP, i2 Technologies, Manugistics, etc.) as supplementary systems to established transaction/ERP systems (SAP 1998). Furthermore, these company-centric packages are currently being extended to provide collaborative planning via the Internet (i.e. SAP SCM's Collaborative Planning). Collaborative planning applications utilize Internet technology (with standardized data formats like XML) to synchronize demand signals and supply chain activities, by allowing supply network partners to view and share common information stored in B2B or even business-to-consumer web sites.

common communication, security and process standards[14], to expand their networking capabilities and transform the nature of their operations. They are pursuing a more narrow control by reconfiguring their supply chains, focusing on core competencies that add value to their supply network, and leveraging skills and information technology to connect and coordinate processes among their trading partners in real time. Such seamless electronic connectivity enables companies to execute networked, cross-enterprise processes and integrate with trading partner operations.

These developments are transforming sequential, enterprise-centric supply chains in which an enterprise drives multiple processes, into synchronized *electronically connected supply networks*, where one process drives more than a single enterprise. E-supply networks may be established either via direct B2B interfaces or via a new breed of Trade Exchanges, or *e-Markets*, which facilitate information sharing, transaction execution and collaborative processes. These predominantly industry-focused e-Markets are most often private, yet there are instances of public, horizontal or consortium-based, i.e. owned by a community of interdependent firms, or even consortiums of competing firms. Such cohesive business networks are confronted by immense challenges; e.g. they need constant communication with customers and suppliers to respond quickly to "pull/push signals" to manage low inventories, adapt quickly and economically to changes in demand/supply, by:

- Taking orders over the web, and provide immediate delivery information (e.g. ATP);
- Offering rich product selection and/or the ability to customize (e.g. we-customizable orders & products);
- Sourcing the order and commit to delivery, immediately, online (e.g. Capable to Promise);
- Service the order online, including changes and inquiries (e.g. order web-flow);
- Deliver product quickly, efficiently, and profitably (integration of logistics & freight information).

---

[14] Examples of such Industry standards: Universal Descriptor Exchange (UDEX) in the Retail and Consumer Packaged Goods (CPG) industry, RosettaNet in High-Tech, and Chemical Industry Data Exchange (CIDX) in Chemicals.

Table 11.1. Summary of objectives of contemporary e-Supply Networks

	Internal Objectives	Downstream Objectives
• Shorten time to market - through collaborative engineering, outsourcing, and contract manufacturing • Provide convenient purchasing via direct web-based sales, online catalogues • Enhance selection through customisation or configurable products • Improve response by order promising, order tracking, event notification and fast delivery.	• Provide visibility of information – inventories, forecasts, orders, plans, engineering changes, kpis. • Synchronise activities – optimised feasible plans, pull-based triggers • Promote responsiveness - reduce time to detect demand, commit, produce, fulfil • Achieve process simplification - by automating routine process steps • Leverage market mechanisms - Aggregate buying power, use auction-based buying/selling via trade exchanges.	• Replace inventory with information (inventory visibility, forecast end-of-chain demand, collaborate with channel / customer), • Shorter planning / replenishment cycles (automated planning process, collaboration with suppliers, rate based planning), • Reduce lead times (through supplier collaboration, "pull" replenishment / VMI and build to order/ postponement), • Improve synchronisation (by generating feasible, optimised plans & schedules, replan when conditions change), • Provide order status and traceability • Use internal and external performance metrics.

These challenges require that partnering companies use *Collaborative Planning and Execution* to reach objectives within the core as well as the up- and downstream domains of the supply network. Application integration together with Internet connectivity enables such real-time communication and advanced planning functionality across multiple enterprises to optimize resource allocation and synchronize information and product flow.

11.2.1 Some Theory – Towards Event-based, Collaborative Business Networks

Beyond the rise and crash of the dot.com era, the study of such inter-organizational business relationships has been central in theories about Business Networks[15] in the last three decades. These research efforts origi-

---

[15] *Networks* are organizational structures in between markets and hierarchies (Hedaa 1997; Ford et al. 1998). The network theories aim to render organizational issues in inter-organizational networks, and focus on strategic positioning or power configurations. Networks typically exist in heterogeneous business-to-business markets, because e.g. trust here is beneficial to all members as it allows the network to define its context and thus its immediate environment (Håkansson and Snehota 1994 in Ford ed. 1997). Network-theory emphasizes the importance of two basic questions: (a) Who does what?, and (b) How are their activities connected? Furthermore, it highlights that companies in general should only perform those activities in which they may perform better than average compared to major competitors in the long run, i.e. focusing upon core

nate from Scandinavia and have been further developed by the Industrial Marketing and Purchasing (IMP) group whose seminal work on networks dates back to 1982. Some of the most noteworthy constructs are the *interaction model*, the *ARA* (Activity links, Resource ties and Actor bonds) model[16], and the *event-based business network*[17]. The first two models study business markets in terms of the nature of buyer-supplier relationships and the fusion of these in industrial networks, modeled as interconnected actors, activities, and resources. Hedaa's *event networks* view interactions as streams of events that ultimately determine effectiveness in networks. Events generated by extensive interactions can reveal exception-handling processes under uncertainty, and provide insights into the dynamics of network evolution[18]. Where strong inter-organizational relationships exist, another type of network that is neither market nor hierarchy, emerges: *network processes*[19]. These *network* or *collaborative processes* represent collaborative arrangements, and rely heavily on information sharing molded by the distribution of power, influence and trust[20]. Better access to material and immaterial resources render some firms more pow-

---

competencies. Where industrial marketing is very much a matter of establishment and development of customer relationships, the network paradigm adds at least three important factors: (1) power, (2) influence and (3) trust.

[16] In general, *actors, activities* and *resources* go into the description of external networks as independent factors (Håkansson and Snehota 1995): a) *Actors* are characterized by their performing of activities and controlling of resources. Actors in an industrial network may be perceived broadly as individual persons, groups in organizational, or organizations. Which actor is going to be at the focus will depend upon the actual context. b) *Activities* are performed by actors when using and transforming resources and considered to be links in longer chains of activities. One such example is the chain of value added in the transformation of raw materials and other inputs into complex products and services. c) *Resources* are controlled by actors and the value of resources is determined by the activities in which they are to be used Examples of resources are technology, finance, capital and personnel (Ford 1997).

[17] See Hedaa and Törnroos (1997).

[18] This is especially articulate in the view of business networks as event networks (Hedaa and Törnroos 1997).

[19] Network processes were mentioned by Easton's chapter in Ford ed. (1997).

[20] According to Thorelli (1993), trust may be viewed as confidence in the relationship, based on awareness of reputation, past performance and reciprocal benefits and demands. Trust determines potential risks and opportunities in network relationships. On the other hand, power and dependency structures often constrain opportunistic behavior, by defining dominant directions of influence.

erful than others, thus stimulating them to pursue network dominance[21]. For example, in supply networks, the obligation to spearhead cooperation often rests with a dominant, highly influential player that defines the ground-rules of collaboration by extending its processes across parts of its web of interactions. In contrast to this extended and enforced cooperation scenario[22], smaller companies, are more predisposed towards loosely coupled collaborative infrastructures[23]. The relative smaller size and consequently lower influence of the network participants create a situation where a company cannot dominate, but rather has to adapt to the network. Configuration of process interactions or links among multiple, equally influential partners are *negotiated* rather than dictated. This in turn requires more introspection of each member's process infrastructure. These issues clearly indicate a rising need to investigate such *e-business* companies that are linked via bilateral and multilateral relationships into loosely coupled process networks, and converge into open Trade Exchanges and/or tighter Collaborative Communities (private e-Markets or Value Chains according to Tapscott et al. 2000). Buyer-seller relations between partners are becoming more opportunistic, endemic and dynamic in nature, while driven by compatible goals. In the face of the rising standardization of communication and data exchange, we do have to reconsider how relationships are evolving within these electronic market networks.

11.2.2 The New Competitive Landscape of e-Supply Networks

Instead of fewer intermediaries in contemporary supply networks, these last 5 years has shown a plethora of new intermediaries entering the buyer-seller relationship. It is evident that companies are able to connect with more partners in business communities, thus creating a multiplicity of network structures on top of each other! As shown in the figure below, complexity increases by additional intermediaries, while flow and ownership of product and information is decoupled. Actually, collaborative planning will take place via a collection of e-Markets, and dedicated B2B links. Some of the new intermediaries entering this interdependent network are shown in the table below.

Thus, e-Markets are effectively functioning as a significant intermediary in the relationship between trading partners. An e-Market is a real-time, marketplace where a buyer can evaluate all the potential suppliers for a

---

[21] From Håkansson and Snehota (1995).
[22] From Browne et al., (1994).
[23] According to Hedberg et al. (1997), this trend is especially prevalent for Scandinavian corporations.

particular product or service. Within a supply network they can be classified as customer facing, e-commerce sites or business-to-consumer exchanges (B2C or e-Commerce), and upstream or downstream, B2B trade exchanges or e-Markets (focusing on corporate customers).

Virtual Manufacturer	Virtual Distributor	Virtual Retailer	Virtual Service Provider
This type of organization does not manufacture anything, nor does it have any plants, but rather, controls product development, marketing, and sales as well as co-ordinate customer service for its products. It hires contract manufacturers and 3PLs and fulfillment service providers to make, assemble, and ship final products to its customers (e.g. Nvidia, parts of Sony-Eriksson's production is outsourced to Flextronics).	This type of organization does not distribute anything and does not have any warehouses. It markets products, takes orders for multiple suppliers, controls marketing and sales, and coordinates order fulfillment. However, it relies on its suppliers to make, assemble, and ship final products directly to its customers (e.g. Ingram Micro).	This type of organization, better known as an Internet retailer, does not own any brick-and-mortar stores. It does, however, merchandise products in virtual stores, namely hosted Websites. The virtual retailer controls order fulfillment and can rely on its own distribution capability or suppliers to ship products directly to customers (e.g. Amazon).	This type of organization does not own any assets, but it does provide SCM services. This includes Lead Logistics Providers that perform logistics management for a company or a Logistics Exchange (LX), which is a trading exchange for procuring and monitoring shipping services (e.g. National Transportation Exchange). Many major 3PLs are developing such "plug-in" services for their customers' private e-Markets.

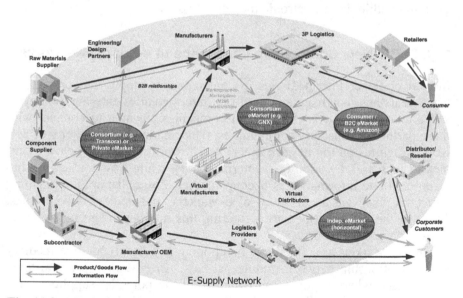

Fig. 11.2. B2B Infrastructures and intermediaries in e-Supply Networks

Table 11.2. Types of e-Markets

Public e-Market or Horizontal Independent e-Market (IeM),	Consortium e-Market (CeM),	Private e-Market of Hub (PeM)
IeM is a many-to-many (m:n) business model, concentrates on the physical transaction – the buyer/seller process. This model pursues to maximize cross-industry or market-based efficiencies in order to achieve cost minimization and asset optimization. Each buyer and seller is but a click away and upon execution of the transaction; they can go their separate ways and may never meet again (i.e. no loyalty). This model is close to the neoclassic characterization of "perfect competition", in that it supports transparent exchange of information such as pricing and availability of all alternative products so that buyers will always be able to make rational decisions. IEs are the natural extension of the Auction model in a B2C or B2B commodity world (e.g. eBay, Free-Markets).	The most potent variant of Public e-Markets has proved to be Consortium e-Market (CeM), which in many respects resembles an electronic version of an industry cartel. Various members of an industry provide the liquidity and momentum in order to achieve industry-specific efficiencies. CeMs concentrate on vertical sourcing and provide a framework for more intense intra-consortium coordination and co-operation (examples: e2open, Covisint, GNX, Transora, Pantellos, etc). While CeMs won't realize the unrealistic return on equity that prompted founder members to invest during the financial bubble, they provide cost and process efficiencies that are not attainable by building and running in-house Private e-Markets. These advantages exist on several levels: infrastructure economies of scale, expanded access to e-business skills, preintegration to efficient trading communities (role- and domain-based), and efficient development and propagation of process and data standards	A Private e-Market or eHub, also called Private e-Market (PeM), is a marketplace established by an entrepreneurial or influential member of a supply network – typically a brand or competence owner. Participation is ensured via cooperative coercion, a new, but very powerful phenomenon that attempts to achieve process and cost efficiencies for a certain subset or segments of an Industry – in some cases it enforces membership (like Daimler-Chrysler's, or WallMart's PeM entry-requirements). In fact, cooperative coercion leads to a tightly-nit, contractual, long-term partnership that pursues collaboration between trading partners. So, PeMs are consolidating pre-established relationships between well-known partners. PTEs are often structured as one-to-many hubs hosted by the supply network host. The initial motivation is procurement cost savings through collaboration, process control, dynamic pricing, plus cycle time and efficiency improvements.

Types of e-Markets

Electronic marketplaces enable companies to efficiently trade and collaborate with their trading partners, and can be described as centralized portals that have either a *vertical* or *horizontal* orientation. *Vertical e-Markets*, service a specific industry segment by delivering one location to transact business. They are "vertical" in the sense that they are channeled to serve specific industries, such as computing, chemicals, steel, and agriculture. Another model is referred to as a *horizontal portal* where, for example, a given process such as procurement or transportation is transacted for several industry segments that share common traits.

Horizontal e-Markets are web sites where buyers and sellers can come together to communicate, share ideas, advertise, bid in auctions, conduct transactions, and manage inventory and fulfillment. They are "horizontal" in the sense that they serve a wide range of diverse industries or address horizontal applications across industries (examples: VerticalNet and TradeOut.com). Another horizontal variant connects customers to a set of suppliers that specialize in a functional supply chain area (e.g. logistics and transportation services).

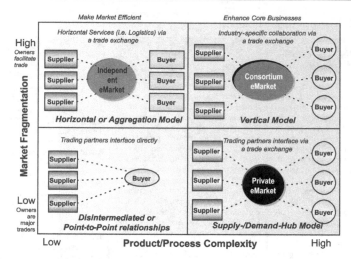

Fig. 11.3. Industry Contingencies and Types of e-Markets

Based on current praxis and research undertaken, we can classify these developments into various types of e-Markets. The real opportunity in e-Markets is the development of collaboration throughout a company's relationship portfolio. Given the ever-increasing need to ensure customer responsiveness and drive industry competitiveness, many companies are currently establishing their own Private e-Markets, or coerced into Consortium e-Markets.

Private or Public e-Markets Developments

After the initial euphoria, the dot-com collapse and the structural changes of the industries (increased ERP penetration), e-Markets are entering a period when technology finally lives up to its promise by creating considerable productivity gains across all industries. The increasing adoption of the Private e-Market model and the progressive adoption of loosely coupled business services made available on demand, lay a renewed foundation for e-Market services, driven by: a) commoditized functionality in supplier and customer self-service portals available as packaged applications; b) improved internal (or A2A) integration with ERP and hosted transaction services; and c) the advent of Business Process Management and Web-services that allow dynamic configuration of loosely coupled services across fragmented industry supply networks (i.e. contract manufacturing and 3rd-party logistics).

These steady, though slow evolutionary developments towards *Collaborative e-Markets* increasingly enable collaborative relationships that share and "jointly derive" planning data, integrate back-end enterprise systems (e.g. ERP), and coordinate supply network activities and resources in real-time among their members. In this context, *collaboration* is the negotiated cooperation between independent companies, exchanging capabilities and constraints to improve collective responsiveness & profitability[24]. On the other hand, Consortium e-Markets that have survived the financial burst (e.g. Pantellos, Elemica, GNX, WWRE), have been developing in a different path than original anticipated, and are currently thriving by offering:

- industry-specific content and collaboration services (via hosted applications), but collaborate on shared business processes, such as settlement and logistics (e.g. Transora has focused on the data synchronization problem for CPG, while GNX offers a broader range of services, some of which stem from partnerships with other service providers).

- e-Markets interconnections (M2M), that enables member companies to get access to a wider selection of services – from product development to financing.

- supplier- and product content and synchronization/translation services (Cross-industry data and process mappings).

- End-to-end supply network collaboration—With broad standardization of basic B2B processes, multi-tier collaborative processes that require Vendor-Managed Inventory (VMI) / Collaborative Planning, Forecasting, and Replenishment (CPFR) are finally getting ready for prime time.

Public or Private or Both!

One further clarification that is currently taking place is the division of labor between the Private and Public e-Markets, in other words, what functionality should be developed in the private domain and what should be subscribed via the public offerings (in e.g. Consortium e-Markets). While this is not an easy question to resolve, most companies have been adopting a portfolio approach to e-market participation, using different models for different business requirements. As depicted in the figure below, horizontal e-Markets are best for settlement and payment (e.g. Swift.com, transactional financial exchange), Consortium e-Markets are best in hosting supplier catalogs (with vertical community content), running integration

---

[24] Specifically, inter-organizational collaboration is defined as a: "process in which organizations exchange information, alter activities, share resources and enhance each others capacity for mutual benefit and a common purpose by sharing risks, responsibilities and rewards" (From: Huxham (1996), Creating Collaborative Advantage, Sage Publishers, London).

hubs[25], and supporting creation of harmonized Product and Supplier Registries (e.g. UCCnet, DUNS). Few CeMs host supply chain event and procurement applications, while Supply Chain Collaboration and Product Development are easiest to deploy within a private e-Market given competitive/sensitivity issues.

Value Added Services	Application Domains	Functionality	Public		Private
			Horizontal eMarket	Consortium eMarket	Private eMarket
	SCEM & Logistics	Supply Chain Event Management	★★	★★★	★★
		Logistics & Transportation Services	★★★	★★	★
	Financials	Settlement, Payment & Clearinghouse	★★★	★★	★
	SRM	Sourcing & RFx/Auctions (Non-strategic)	★★	★★★	★
		Procurement (Indirect / Non-strategic)	★★	★★★	★
		Sourcing & RFx Services (Direct / Strategic)	★	★★	★★★
Business Applications		Procurement (Direct / Strategic)	★	★★	★★★
	PLM	Asset Maintainance & Management	★	★	★★★
		Collaborative Design & Engineering	★	★★	★★★
	SCM	Collaborative Project Management	★	★★	★★★
		Supply Chain Collaboration (e.g. CPFR)	★	★★	★★★
	Content Mgmt	Order Fulfilment	★	★★	★★★
Content & Data Management		Catalogs & Content Services (Non-strategic)	★★★	★★	★
		Catalogs & Content Services (Strategic)	★	★★★	★★
	CRM	Community Content & Industry Registry	★★	★★★	★
	BI	Customer Service and Support	★★	★★	★★★
	Middleware	eAnalytics (Performance Monitoring+Reporting)	★	★★	★★★
		B2B Transactional Integration/Mapping Services	★★	★★★	★
IT Integration		B2B Document Routing & Security Services	★	★★★	★★
		B2B EDI VAN Replacement & Registry Services	★★	★★★	★

Fig. 11.4. Creating e-Markets Functionality Portfolios - Optimizing Utility & Execution

Distinguishing between e-Market value-added services require that companies evaluate the three major inter-organizational issues:

1. IT Integration – the common logic here is to avoid setup and operational cost of a private e-Market infrastructure for non-core/-strategic processes, which will enable the company to link to business partners more cost-efficiently to exchange business documents.
2. Content & Data Management – there is a clear trend towards increasing use of Punch-out Catalogs (roundtrips) rather than Local Catalogs. Many companies realize the difficulty of enabling and maintaining local supplier content (e.g. Shell), while few suppliers have created ad-

---

[25] Integration Hubs use different data formats and protocols like EDI and XML (xCBL, CIDX, CPFR, RosettaNet) to map and translates protocols, standards, and data formats between multiple firms' systems (current examples are Transora & Elemica).

vanced sales catalogs. Use CeMs to reduce costs of Supplier Activation and Content Management for non-core/-strategic content compared to pure in-house deployment. Further companies should exploit economies of scale in information provisioning (e.g. non-sensitive supply chain and Industry information) for enablement of collaborative processes via CeMs.

3. Hosted Business Applications – running multi-firm applications is cumbersome! CeMs can enable collaborative multi-firm capabilities by linking information and processes (i.e. supply chain visibility, notifications, and information pooling) via Web Services.

One of the most potent development in the wake of the Private and public e-Markets was the creation of standardized process definitions and data interfaces or relationships. All that enables individual companies (or communities) to create a transactional *hub* for the facilitation of information, business transactions and collaborative processes. So, instead of creating customized links to the company's strategic partners, the marketplace becomes the central conduit of most business relationships.

Enabling Integration – From EDI to B2B XML

On the pre-Internet era most business focused exclusively on internal optimization; inter-organizational, B2B relationships were handled on a one-on-one buyer-seller basis without any benefits or synergy being derived from pooling any processes or transactions across the supply network. In this "Old World" each individual connection or link to a business partner needed integration (via EDI, Edifact, FAX, etc.), and customization of back-end systems that in turn required constant maintenance. Prohibitive costs related to the setup of such *one-to-one* (1:1) relationships, left many companies out of the integration loop, thus technology did not lead to any significant benefits.

With the emergence of e-Commence and e-Business, numerous e-Markets sprouted, which promoted the realization of e-Supply Networks, characterized by a virtual number of potential trading partners coming together to share information, transact business, and collaborate. While EDI-based interconnections have proved too costly and inflexible to be the integrating vehicle for the Digital Economy, e-Markets enhanced by new developments in process and data-standards (e.g. XML, Java) are squeezing transaction costs further down. Collaborative planning & execution can take place via B2B or B2M2B scenarios. In *Point-to-Point, bilateral* (1:1) relationships imply that focal company have to establish formalized relationships with each partner (either via EDI or Internet), where as *Marketplace* (B2B or B2M2B), or *multilateral* relationships, imply the creation of one, broadband interface to an e-Market (public or private), where partners interact in a one-to-many or a many-to-many collaborative planning environment.

Hence, companies are able to create more dense interactions, consisting of interrelationships between process activities, participating actors and applications. Consequently, the nature of relationships between actors in different companies is also altered as more intra-company processes are extended beyond the boundaries of the firm. Collaborative processes running via an e-Market shift the decision-making process from within the company to the relationship between companies, where decisions are derived in a *joint fashion*.

11.3 Intercompany Relationships in e-Supply Networks

Collaborative Processes use Internet connectivity and standards to enable real-time communication and planning functionality across multiple enterprises to synchronize information and product flow, in order to optimize resource allocation and minimize costs. It may help to reduce inventory across enterprises, maximize network capacity utilization, improve service levels, shorten planning cycles, pull rather than push products, identify critical supply issues, and introduce sophistication and clarity into the process. Collaboration with upstream and downstream partners can take many forms, including mass customization to joint product development, shared forecasts, and co-location or other managed inventory practices, yet it requires that the internal processes are in place (see figure below).

As can be seen, this changes the nature of the relationship and hence the transaction between trading partners. Instead of buyer/seller relationships we have a range of relationships. Collaborative Planning may take place either via *B2B* or *B2M2B collaborative scenarios*.

Companies may view collaboration as a means to synchronize supply chain operations, particularly with regard to strategic, tactical, and operational planning activities. Collaboration may involve optimizing and integrating various planning processes in the supply network, like: Sales & Ops-, Demand-, Capacity- Supply-, Production-, Product Lifecycle-, Category-, Transport-, and Merchandise Planning. Many companies are currently experimenting with systems technology to speed up operational and financial transactions with trading partners by using EDI and more recently, e-Markets. This coincides with the increasing automation of internal processes, which is necessary to conduct B2B commerce. Furthermore, many internal production and distribution processes like MPS, MRP, DRP are moving outside the boundaries of the firm (e.g. Vendor Managed Inventory has changed the replenishment process). The goal of these optimization and integration efforts, is to provide functionality, such as:

- Real time communication, including business logic, where each event is monitored by alerting systems for real time transactional data and decision support information about customers and orders.
- Shared resource allocation, document generation, and profitability monitoring.
- Deliver to promise, where rates and routes are chosen accurately and dynamically, giving delivery time in hours & minutes.

11.3.1 Relationships through e-Markets

e-Market-enabled business relationships often involve the automation of various aspects within a buyer/seller or trading relationship. Contemporary implementations of inter-organizational partnerships focus on enabling B2B planning (especially via Consortium or Private e-Markets). Remark that collaborative relationships involve some sort of synchronized planning plan execution. While there are myriad aspects within a collaborative planning relationship among trading partners in a e-Supply Network, three broad e-Market-enabled relationship categories have been identified: a) *Information-Sharing*, b) *Integrative*, and c) *Collaborative*.

Information-Sharing Relationships via e-Markets

Historically, little information has been electronically shared among trading partners. The first collaborative planning relationship follows the automation of buyer-seller EDI-based transactions (mostly procurement or replenishment), and involves *information sharing* or data exchange. This involves at least one of the following arrangements: a) The partners are given access to an area of an e-Market that has the shared information in it, or b) One partner transmits shared information to the other partner. For example, Web-based collaborative planning books allow buyers to electronically view planning information (e.g. demand forecasts). From a buyer-side, automation has focused on electronically providing forecast needs. In this type of relationship, information ancillary to actual plans is shared only on an FYI basis.

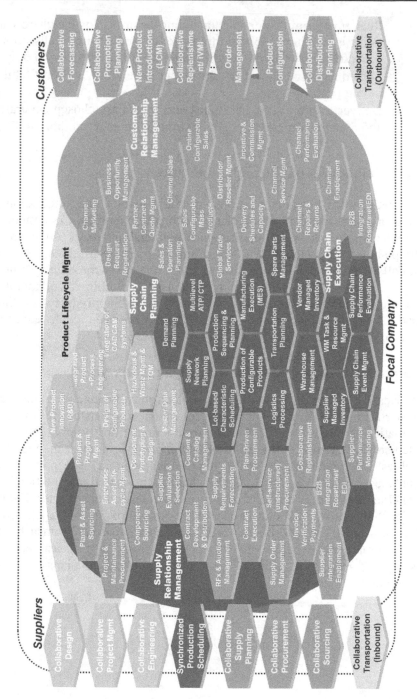

Fig. 11.5. From Internal to Collaborative Process Interfaces – A Process Mosaic

Table 11.3. Summary of objectives and challenges facing a company in e-Supply Networks

Information-Sharing	Integrative or Trans-actional	Collaborative
Information-Sharing relationships mean that partners are given access to an area of an e-Market that has the shared information in it, or one partner transmits shared information to the other partner	Integrative relationships support information-sharing and Computer-to-Computer transmission of fixed structure transactional information.	Collaborative planning relationships facilitate collaborative relationships, where many-to-many information is not just exchanged and transmitted, but is jointly developed by the buyer and seller.
• Most B2B transactions are taking place outside the marketplace (via email, fax and mail) • Supports synchronized, but independent planning and forecasting (one-to-many, many-to-many) • **Minimum support of integrated execution – such e-Markets function as middleware and message brokers** • **Information sharing relationships differ from collaborative relationships primarily in that information is sent on an FYI basis**	• Rich information exchange and event notification (one-to-many, many-to-many) • Most transactions between backend systems (ERP) are transmitted via the marketplace • No support of synchronized planning – planning is still completed within each partner • **Supports synchronized execution of routine transactions (i.e. Order fulfillment, Replenishment)** • **These activities involve information notifying the buyer and seller that a purchase is taking place and that funds need to be exchanged**	• All collaborative relationships involve some sort of joint planning and plan execution. • Rich information exchange and exceptions/alert notification • Most transactions between backend systems (ERP) are transmitted via the marketplace (one-to-many, many-to-many) • Supports joint synchronized planning and synchronized execution of routine transactions (i.e. Order fulfillment, Replenishment) • Most routing processes are driven by real time exception handling

Buyer or seller can share various types of information, either before or after a purchase is made. This information may involve the seller's offerings or the buyer's future needs. Information-sharing relationships differ from collaborative relationships primarily in that information is sent on an FYI basis. The recipient is using the data as-is and is not providing feedback. An important exception is that this may differ in Private e-Markets or e-Hubs. Here a partner's internal SCM system may publish information and expect feedback. Nonetheless, this information is helpful in improving supply chain performance. Information-sharing arrangements electronically support both supply chain planning and execution, thereby presenting the potential to improve overall network performance. Relative to planning, these arrangements only support independent planning done by each participant, rather than joint planning. Forecasts developed independently from trading partners, and pushed to upstream. However, sharing helps to ensure that trading partners' plans are as synchronized as possible, which in turn effectively reduces uncertainty in their supply and demand situation. Rather than having to predict or forecast a partner's activities, information sharing ensures that the parties are knowledgeable about each other's activities.

Automation of transactions is not taking place within an Information-Sharing e-Market. They may involve activities conducted to execute the buyer's purchase of a commodity. These activities involve information notifying the buyer and seller that a purchase is taking place and that funds need to be exchanged. Thus automation focuses on using EDI to electronically send purchase orders and invoices, and to transfer funds. The only information that can be transmitted in this type of relationship is that needed to execute a purchase. So, to summarize, this relationship supports synchronized, but independent planning and forecasting (one-to-many, many-to-many), provides minimum support of integrated execution, while most B2B transactions are taking place outside the marketplace (via email, fax and mail).

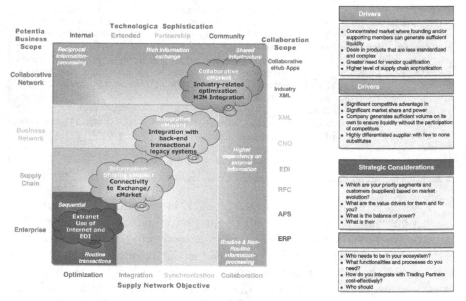

Fig.11.6. e-Market types based on Collaborative Relationships

Integrative Relationships via e-Markets

While information-sharing relationships enable supply chain synchronization, they do little to reduce the uncertainty faced by trading partners in determining future demand, and do not grant the opportunity for the other partner to provide his or her own insight and knowledge of customer needs or other market opportunities. In addition, there is little opportunity to work together on matching supply with anticipated customer demand. To further enhance a buyer-seller relationship some progressive companies are moving toward *collaborative relationships*, in which they are "working jointly with others, especially in an intellectual endeavor." Collaborative efforts enable trading partners to work together to better understand future demand and to put plans in place to satisfy it profitably.

An integrative e-Market facilitates collaborative relationships, where many-to-many information is not just exchanged and transmitted, but is also jointly developed by the buyer with the seller. For example, in the case of working collaboratively on customer requirements, trading partners might work together on new product designs and customer demand forecasts. Generally this information deals with future product plans and needs. Much like an information-sharing relationship, related information to an actual transaction is shared in a collaborative environment. Yet, either party may alter joint plans. A trading partnership between a particular buyer and seller could be based on various exchange modes. That is, some information may be exchanged on a transactional basis, some on an information-sharing basis, and some on a collaborative basis. These type collaborative planning relationships require that transactions are transmitted via the e-Market. That means that a shared repository exists that facilitates and integrates both data and transactions between people and systems. To summarize, this e-Market type facilitates rich information exchange and event notification (in one-to-many, many-to-many scenarios), it integrates most transactions between backend systems (ERP), and supports synchronized execution of routine transactions (i.e. Order fulfillment, Replenishment), and Computer-to-Computer transmission of fixed structure transactional information. These activities involve information notifying the buyer and seller that a purchase is taking place and that funds need to be exchanged. However, Integrative e-Markets still do not support synchronized planning via the shared marketplace – final planning is still completed within each partner's planning system (e.g. APS).

Collaborative Relationships via e-Markets

The two previous types of collaborative planning relationships are nothing more than facilitators of communication, though probably the first to be implemented. Lacking well-defined business processes and industry wide product standards, many vertical e-Markets will provide just the basic infrastructure for linking companies together. How? By establishing XML-based standards and aggregating data across participants. Rather than building a proprietary B2B infrastructures, participants expect these shared collaboration environments to:

1. *Reduce supplier integration costs* by allowing firms to integrate to multiple customers through a single e-Market, substantially cutting down on the slew of integration projects.
2. *Minimize investment expense* by allowing firms to share the development and ongoing maintenance costs, rather than creating redundant systems and capabilities.
3. *Optimize industry wide capacity* by pulling together supply chain information from many firms, and offering a consolidated picture of industry capacity and market demand in order to optimize inter-enterprise production.

Thus, *information-sharing* and *integrative collaborative e-Markets* simplify buyer/seller integration through a single communication & coordination venue. Nevertheless, both models require that most members still own and maintain elaborate internal SCM systems. Another model is slowly emerging that will probably in some industries overtake the other two flavors. The *Collaborative e-Market* will play the part of *industry optimizers*, by actively coordinating entire supply networks. These full-featured sites will monitor cross-enterprise demand and capacity to fulfill manufacturers' needs with optimal supplier capacity. Participants will directly own them through Private-, Public, or Consortium-oriented constellations, manage these venues and support them primarily through membership and service fees. Manufacturers and suppliers that connect into these hubs will pay for SCM system and Event-/Exception Management System services through an ASP-like model of subscription fees along with à la carte payments for additional services. That makes this model a favorite Private e-Market/ eHub configuration among large, dominating players (like Dell, IBM) that want to consolidate their relationship portfolio, but it will also be appropriate among fragmented industries, where many small partners will join forces to create Collaborative Communities.

These complex e-Markets aggregate demand & supply, match buyers & sellers, consolidate capacity, monitor multi-level performance and notify changes real-time based on internal exception management rule-engines. They are able to share data between ERP systems, reserve or route ATP requests, and conduct N-tier mapping (multi-company BOM explosions and dependent requirements). Supply network optimization will take place via Collaborative SCM/SRM/PLM tools executing from within the e-Market, which will then transmit rich planning information and exceptions/alert notification to the members.

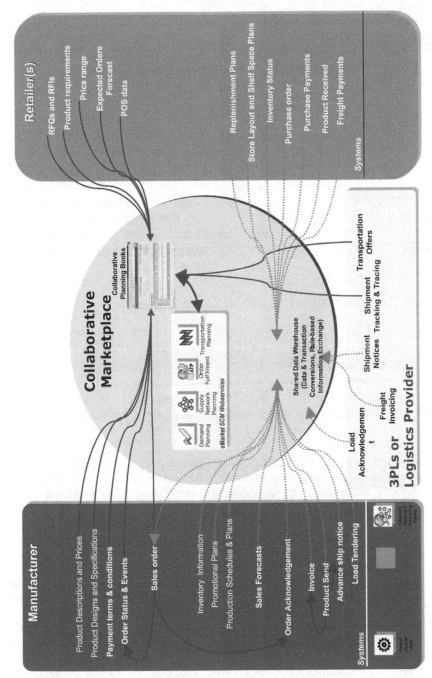

Fig. 11.7. Collaborative e-Market – Enabling Joint APS Execution via hosted SCM applications and web services

Most transactions between back-end systems (ERP) will be transmitted via the e-Market (in both one-to-many, many-to-many modes). Collaborative e-Markets support joint synchronized planning and synchronized execution of routine transactions (i.e. Order fulfillment, Sourcing, Replenishment), while more advanced versions will deliver a full range of transactional relationships, like collaborative production scheduling, and collaborative product development. To summarize, these e-Markets deliver an extensive collaborative platform to jointly plan & execute a wide range of activities (i.e. Design & Engineering, Sourcing, Manufacturing, Sales, Distribution & Transportation). Such collaboration will ensure a) visibility by real-time communication in the supply network, b) performance transparency, and c) responsiveness, by reducing time to detect demand, commit, produce, and fulfill buyer demands. It may sound like rocket science, but many companies (e.g. Wal-Mart, Cisco) are currently evolving their Private e-Market to support such collaborative relationships.

11.4 Collaborative Processes in e-Markets

Within a *e-Market* that delivers/supports collaborative planning, the standard supply chain processes – *plan, source, make, deliver* – as defined by the Supply-Chain Operations Reference-model (*SCOR*), a standard process reference model[26], developed and endorsed by the Supply-Chain Council[27], are transformed into their collaborative counterparts. SCOR outlines the key inter-linked supply chain processes and their component sub proc-

---

[26] A process reference model describes, characterizes and evaluates a complex management process. Such a model builds on the concepts of BPR, benchmarking and process measurement, by integrating these techniques into a cross-functional framework. Once a complex management process has been "captured" in a process reference model, it can be described unambiguously, communicated consistently, and redesigned to achieve competitive advantage. In addition, given the use of standard measurements for process elements and activities, the process itself can be measured, managed and controlled, and it may be refined to meet a specific purpose. Process Reference Models accommodate a number of constructs by providing a balanced horizontal (cross-process) and vertical (hierarchical) view, they are designed to be (re)configurable, and are most often used to represent many different configurations of a similar process as an aggregate of a series of hierarchical process models.

[27] The Supply-Chain Council (SCC) was organized in 1996 by Pittiglio Rabin Todd & McGrath (PRTM) and Advanced Manufacturing Research (AMR), and initially included 69 voluntary member companies (today over 450).

esses, which may assist companies in evaluating their supply chain performance, identifying weak areas, and developing improvement solutions.

Rather than the chain oriented metaphor used to depict the 4 processes, e-Markets do not require bilateral or point-to-point relationships, but support multilateral interfaces between its members. From a process and applications support perspective, the requirement for front facing customer processes to be integrated with back-end transactional processes becomes cross-company in scope. Companies will have to bridge or supplant information from their internal Enterprise Resource Planning (ERP), APS, CRM, and legacy applications to one or many e-Markets, either initially via a Web browser, or eventually via system-to-system integration. The building blocks of such communication are a) common business documents and transactions, and b) common semantics, taxonomies and standards (both data and process).

In the figure above we show some of the collaborative processes that may take place within an e-Market. As noted by the SCOR framework, the planning process spans all other processes, making it the fundamental linkage of manufacturing execution, sourcing, delivery, monitoring, and control. As you can see, we have categorized most collaborative planning processes under these headings. Also, operational processes taking place either outside or through the e-Market, feed information to the collaborative planning processes. Collaborative planning may then take place in either external fashion - meaning within each company in information-sharing and integrative e-Markets, or within the actual e-Market (in a Collaborative e-Market).

The CPFR process model

As we have seen, SCOR is an excellent model to provide the necessary overview and classification of collaborative arrangements. The next vexing question is "how" are we going to initiate and establish such collaborative processes. This is the realm of the CPFR model (*Collaborative Planning, Forecasting, and Replenishment*), which according to the Voluntary Interindustry Commerce Standards, "... *is a business process model for value chain partners to coordinate plans in order to reduce variance between supply and demand*" (see www.cpfr.org). CPFR is a business process model that companies use to optimize supply chain activities such as Vendor Managed Inventory (VMI) by leveraging the Internet and EDI to radically reduce inventories and expenses while improving customer service. Historically, CPFR grew out of the retail consumer goods industry. We focus on this model because it is the most widespread accepted, piloted, studied and enabled (by SCM software).

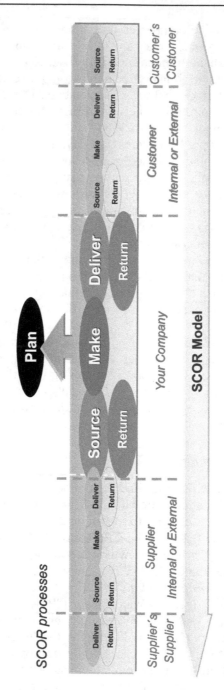

Fig. 11.8. The Supply Chain Council's Supply Chain Reference process model (Rel. 5.0)

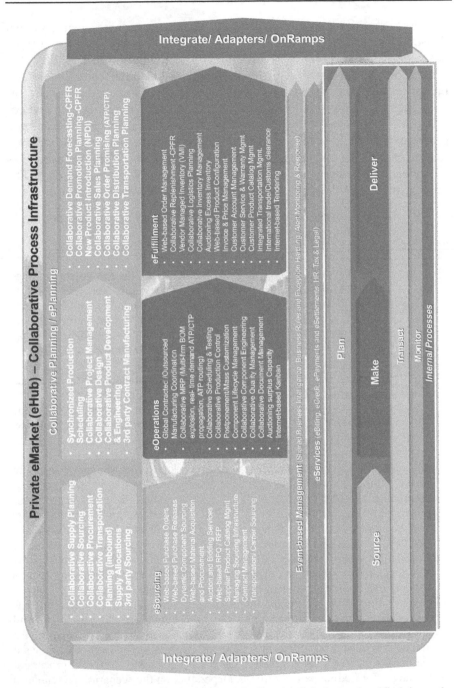

Fig. 11.9. Collaborative Processes in Advanced e-Markets classified into the SCOR processes

CPFR provides a set of guidelines on *how* companies can establish dense, collaborative partnerships within a supply network[28]. From a business process standpoint, CPFR defines how retailers and suppliers can synchronize their different planning functions. Retailers are focused on predicting customer reaction to promotions, competitors, and product category changes, while suppliers usually concentrate on managing the level of inventory at distribution centers. While the retailer's objective is to keep products in stock in stores, the supplier's objective is to create the most efficient production and replenishment process possible. These differences are reflected in each party's sales and order forecasting processes [29]. The guiding principles developed for CPFR out of the VMI best practices are:

–   The trading partner framework and operating process definition focus on customers[30].

---

[28]   Today, more than 30 US companies are part of the CPFR committee, amongst then Wal-Mart, Kmart, Schmuck, Wegmans, SuperValue, Butt, Target on the retail side as well as manufacturers such as Procter& Gamble, Sara Lee, Levis, Nabisco, Kimberly Clark, Kodak Heineken. From the IT side partners include Sun, Hewlett Packard, IBM and SAP. The concept is also extending to Europe and currently Procter & Gamble is running pilots in 4 countries with 4 leading retailers.

[29]   Sales (Consumer Demand) Forecast Comparisons Retailers produce very detailed sales forecasts, often including weekly (or even daily) store-level demand per SKU. Suppliers may also gather a great deal of intelligence about what sold from a syndicated data source (typically IRI or Nielsen), but they usually create only market- or account-level forecasts. The CPFR solution aggregates the more detailed sales forecasts from the retailer and compares the total with the supplier's number. Order Forecast (Replenishment Plan) Comparisons Often, retailers do not produce an order forecast at all. When retailers do produce an order forecast, it may include only base demand. Many handle promotional orders through a totally different process, tools, and personnel. Suppliers, therefore, don't often get an integrated view of the retailer's demand. A CPFR solution can improve this situation by providing a forum where replenishment order forecasts and promotional orders can be brought together and compared in full. It can also give the retailer better visibility to how the supplier makes changes to their order forecasts to meet demand.

[30]   One key finding that has come out of the programs is that no single business process fits all trading partners or all situations between trading partners. Trading partners have different competencies based on their strategies and investments. They also have different sources of information and different views of the market place. CPFR is structured as a set of scenarios or CPFR process alternatives for trading partners to use. Depending on the scenario, the retailer or the manufacturer may be responsible for specific parts of the collaboration process.

- Trading partners manage the development of a single shared forecast of customer demand that drives planning across the value chain[31].
- Trading partners jointly commit to the shared forecast through risk sharing in the removal of supply process constraints[32].

Applicability of the CPFR business model

According to experiences gained by the case companies that have already implemented CPFR based collaborative processes, CPFR does not itself fit all B2B collaborative needs. Products that are commodity-based, have many alternative sources of supply, are undifferentiated, or where price is the primary driver for acquisition, a many-to-many e-Market model makes more sense. This is because a generally public/consortium e-Market that focuses on transaction cost reduction works. Buyer and seller are both motivated to reduce the cost of doing business – with any buyer or seller.

---

[31] A single shared forecast is developed which is then shared across the entire supply chain, to ensure that both retailers, wholesalers, manufacturers and suppliers work towards a common goal. Retailers and manufacturers have different views of the marketplace. Retailers see and interact with the end consumer in person and infer consumer behavior using POS data. They also see a range of manufacturers, their product offerings, and their plans for marketing those products. Manufacturers see a range of retailers and their merchandising plans. They can also monitor consumer activity, with some delays, through syndicated data. Given these different views, the trading partners can improve their demand planning capabilities through an interactive exchange of data and business intelligence without breaching confidences. The end result is a single shared forecast of consumer demand at the point of sale. This single shared-demand plan can then become the foundation for all internal planning activities related to that product for the retailer and the manufacturer, all the way to the manufacturers suppliers. In other words, this single shared forecast is the basis for the synchronization of the extended supply chain.

[32] The value of having a single demand plan, if nothing else changes, would be to better co-ordinate value-chain process activities. This co-ordination would yield significant, but not dramatic benefits. Dramatic benefits come from using the demand plan to affect the significant constraints inhibiting supply-process performance. An example of a significant constraint would be manufacturing flexibility. Most manufacturers hold finished goods inventory in sufficient quantities to meet retail demand. Manufacturing capacity is not used because the retailers' normally short order-cycle times are inconsistent with longer manufacturing cycle times. By extending the retailers' order cycle and thus making it consistent with the manufacturing cycle, production could move to a "make-to-order" process for some products.

CPFR better fits any need where these characteristics are not apparent. CPFR is more applicable where customer service and buyer and seller agree to forgo the benefits of a short-term (i.e. price deal) for the greater mutual benefit of a longer-term relationship. Thus, electronically driven CPFR processes are most appropriate where service and product, not price differentiation, is the factor in the buying decision. CPFR works best where the focus is on long-term relationships for highly differentiated products with limited sources of supply (see *CPFR Roadmap*, 1999).

11.4.1 Collaborative Planning in e-Markets

In accordance with CPFR pilot results the major collaboration opportunity areas are in demand planning and inventory replenishment. Yet, this is only the beginning. Upcoming e-Markets, whether being information sharing, integrative or even collaborative, will implement standardized data and process that will support a range of processes as depicted in the figure below. In the figure below we have depicted the major initiatives within each process domain (e.g. CPFR).

Multilateral relationships among trading partners within an e-Market often differ depending on the companies involved. In general, collaborative relationships dependent upon the specific buyer and seller involved. It is highly unlikely that all trading partners will have the same relationships with the buyer or seller. There will always be favored suppliers and customers with different collaborating capabilities. Additionally, electronic collaboration will differ substantially by a trading partner's role within the supply network, depending on whether it is a manufacturer, distributor-wholesaler, retailer, or 3PL provider. The most important collaboration opportunity areas will vary along a supply network and are likely to result in three major, clusters of buyer-seller collaborative planning processes:

1. Manufacturer with its suppliers (including tier supplier with its suppliers)
2. Manufacturer with its customers (e.g., wholesale-distributors and retailers)
3. Companies with their 3rd Party Logistics (3PL) providers

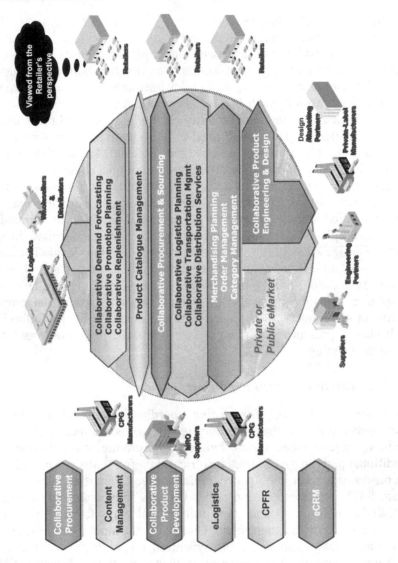

Fig. 11.10. Central Collaborative Processes in a Consortium or Private Retail e-Market

11.4.2 Manufacturer-Customer Collaborative Planning Processes

For finished-/brand goods manufacturers and their customers (such as wholesale-distributors and retailers) the major collaboration opportunities lie in demand planning and inventory replenishment. By collaborating and synchronizing sales forecasts these supply networks attempt to jointly evaluate customer demand at the point of consumption, i.e. retail store shelves. Once established, a replenishment plan that meets the anticipated demand will be mutually agreed upon. Coordinating both the demand and replenishment plans will help ensure that customer requirements are met in an optimized fashion. Such collaboration requires that the partners cooperate electronically to share and modify each other's demand plans and forecasts. Each trading partner will need to understand the other's promotional plans and the plan's impact on customer demand. Within this context, it will be important to electronically share promotional calendars that include anticipated marketing actions designed to stimulate customer demand pricing actions, customer promotions (e.g., coupons), advertising plans, new product introductions, assortment plans, etc. In addition to demand forecasts and replenishment plans, a manufacturer and retailer may collaboratively manage a category of products, possibly at store level. This will require that they electronically collaborate on store layout and shelf space plans. In addition, POS (point-of-sales) data involving store-level demographic information must also be shared, to jointly assess the proper assortment of products to be placed within each store. In the following we will shortly present the 3 processes of demand-, promotion- and replenishment planning.

Collaborative Demand Planning

Collaborative Demand Forecasting coordinates demand and replenishment plans to ensure that consumer requirements are met in an optimized fashion, by jointly developing forecasts and promotional calendars. While traditional planning/APS, uses historical data for statistical modeling, and incorporates market intelligence, collaborative forecasting uses POS data - store level consumer demand rather than DC replenishment, agreed consensus-based forecasts and joint promotional plans to reach and optimal forecast and replenishment plan. Thus, relevant input from business partners can be taken into account to synchronize planning across the network to generate optimized plans based on data from the e-Market. Collaborative forecasting may be undertaken by an e-Market designed to:

- Enable the exchange of appropriate up-to-date planning information with partners
- Allow easy access using the Internet to read and change data (via planning books)
- Restrict user access to authorized data and activities (via a Data Warehouse)

- Support consensus planning process (through shared planning books)
- Support exception-based management (though alert notification via email)
- Generate 'one number' for forecast across the supply network.

The most widespread collaborative process currently, *Collaborative Demand Planning* between manufacturers and their distributors/customers, allows both partners to streamline their work processes and ultimately benefit from a more accurate forecast, better market transparency, greater stability, reduced inventory and better communication. Buyer and seller develop a single forecast and update it regularly based on information shared over the e-Market. It is a B2B workflow, with data exchanged dynamically, designed to increase in-stock customer stock while cutting inventory.

The basic process of the CPFR model consists of 9 steps defined by (see detailed view in the figure):

Step 1 – *Front-end agreement*: Participating companies identify executive sponsors, agree to confidentiality and dispute resolution processes, develop a scorecard to track key supply chain metrics relative to success criteria, and establish any financial incentives or penalties.

Step 2 – *Joint business plan*: The project teams develop plans for promotions, inventory policy changes, store openings/closings, and product changes for each product category.

Steps 3-5 – *Sales forecast collaboration*: Buyers/Retailers and suppliers share customer demand forecasts, and identify exceptions that occur when partners' plans do not match, or change dramatically. They resolve exceptions by determining causal factors, adjusting plans where necessary. This is achieved by comparing current measured values such as stock levels in each store adjusted for changes such as promotions against the agreed-upon exception criteria (in-stock level, forecast accuracy targets).

Steps 6-8 – *Order forecast collaboration*: Develop a single order forecast that time-phases the sales forecast while meeting the business plans inventory and service objectives, and accommodating capacity constraints for manufacturing, shipping, and more. Identify and resolve exceptions to the forecast, particularly those involving the manufacturers constraints in delivering specified volumes, creating an interactive loop for revising orders.

Step 9 – *Order generation/delivery execution*: Generate orders based on the constrained order forecast. The near-term orders are fixed while the long-term ones are used for planning. Results data

(POS, orders, shipments, on-hand inventory) is shared, and forecast accuracy problems, overstock/understock conditions, and execution issues are identified and resolved.

Collaborative Promotion Planning

Collaborative Promotion Planning between distributors and their customers allows these supply network partners to streamline their work processes and create a more accurate plan; e.g. the distributor's promotion planning data created in Collaborative Forecasting is accessible to external partners via the e-Market, who can then decide to participate in a planned promotion. In Promotion Planning via an e-Market, the external partner in the collaborative planning process accepts or rejects a promotion offered by the distributor or manufacturer. By accessing the shared planning books or data warehouse of the e-Market, the external partner can: display a list of promotions, display detailed information such as periods and quantities, accept or deny the offer, attach a note to a promotion plan.

The brand manager initiating the scenario shown in the figure above identifies an area that requires incremental promotional activity. Once the target and the area have been identified, the promotion is established and passed on to the field organization. The account manager receives notification and is instructed to develop promotion plans for the company's accounts. Having set the objectives of the promotion, initial volume lift factors, and promotion elements such as media and trade support, together with the allowances and funds, the account manager is able to develop the events. These events are then are presented to the retailer for approval and, if they are accepted, the two parties work together to agree on the details of the promotion and volume estimates. When the final promotion plan is in place, the account manager orders additional promotional materials, such as display pieces, and passes the promotion details on to the supply chain manager. The supply chain manager integrates the details into the plan and the promotion is run. To complete the cycle, both the retailer and manufacturer are able to evaluate the effectiveness of the promotion and use what they have learned for future planning. By combining promotions and demand forecasting plus information regarding new product introductions, partners are able to streamline the demand signal and achieve substantial benefits.

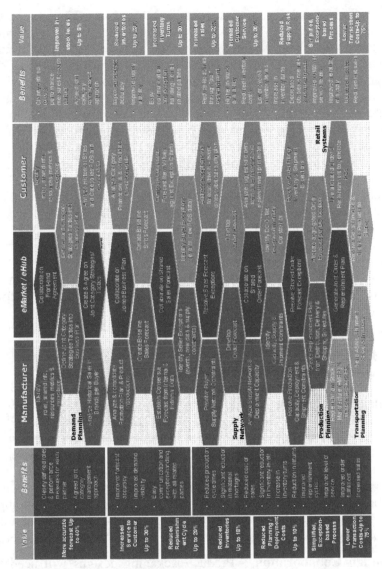

Fig. 11.11. CPFR – From Collaborative Demand Planning to Collaborative Replenishment, Enabling Applications and Benefits

Collaborative Replenishment

Collaborative Replenishment takes over after completion of the aforementioned processes - collaborative forecasting and collaborative promotion planning. Via e-Markets partners have the opportunity of automating large parts of the replenishment transactions via Internet-based Vendor Managed Inventory (iVMI), which utilizes XML formatting to exchange information between systems. VMI is a service provided by a supplier for its customers whereby the supplier takes on the task of requirements planning for its own products within the retail company. For VMI to work, the supplier not only must be able to track the amount of its products stocked at the customer site, it must also take into account the customers sales forecasts. Making VMI possible via Internet provides small retailers with an economical alternative to participating in supply chain planning. It also allows the retailer to maintain control over the data it is sending to the supplier and change it if necessary. To achieve their goals, participants will be able to access the Supply Network Planning data through Internet planning books residing in the e-Market. To summarize, automated replenishment or iVMI is a strategy aimed at enhancing the efficiency of a partnership. VMI is only possible if the manufacturer has information on sales figures/current stock levels -> retailer needs to transfer them. Typically, the result is that the manufacturer can forecast future sales and replenish retailer inventory more efficiently, which also means that it is easier for him to plan production.

11.4.3 Manufacturer-Supplier Collaborative Planning Processes

The major benefits that a manufacturer will get from collaborating with its suppliers include new *product development* and *synchronized production scheduling*. The latter can be segmented into *collaborative supplier planning* (for strategic & tactical decisions), *collaborative procurement* (for operational day-to-day requirements), and *collaborative production execution* (primarily for outsourced production, or subcontracted production). Collaborative product development will yield benefits by helping the manufacturer to develop stronger products more efficiently. There are several major opportunity areas within collaborative product development:

- **Design Collaboration** – Product/packaging designs will need to be electronically shared and modified--possibly using CAD files.
- **Product-Costing Information** – Costing data will need to be shared and mutually established to help ensure that target product costs are achieved.
- **Subcontracting Relationships** – Contract terms and conditions will need to be jointly established and contracts electronically passed back and forth for modification and approval.

In a similar fashion, coordinating or synchronizing all tier-supplier production schedules will help ensure that future material needs are satisfied, resulting in improved order fulfillment. This is often realized by electronically sharing schedules with suppliers, allowing them to provide feedback and make changes based on whether or not material needs can be met. This type of collaboration also includes visibility into the raw material, WIP, and FG inventories of all suppliers to help ensure synchronized realistic production schedules.

Collaborative Supplier Planning

Collaborative Supplier Planning, enables suppliers to access to production plans as well as dependent requirements, which enables them to use consumer demand customer inventory levels to fine-tune replenishment; materials requirements are shared at an early stage between manufacturers and suppliers so that all parties involved can adjust their supply and production plans; e.g., if the delivery of the dependent requirements can't be made in time, an alternative date can be suggested. The goal of this process is to help enterprises carry out collaborative supply chain planning activities with their business partners. Thus, relevant input from business partners can be taken into account to synchronize planning across the network, in order to generate optimized plans based on data from the supply network. Enterprises can now focus on enhancing customer value by enabling true business collaboration among business partners in their networks. Collaborative supplier planning may be conducted or executed by an e-Market designed to:

- Enable the exchange of appropriate and up-to-date required planning information with business partners
- Restrict user access to authorized data and activities
- Support a consensus planning process and exception-based management
- Generate 'one number' for supply chain planning across networks

During the course of Collaborative Supply Planning the manufacturer and supplier exchange information on the material requirements of the manufacturer, and they collaborate on exceptions. This type of collaboration enables both the manufacturer and supplier to create more accurate supply network and production plans. The plans can be updated regularly based on information shared over the Internet. This is a business-to-business workflow, with data exchanged dynamically, which is designed to decrease inventory. The basic process consists of seven steps:

1. Both partners agree on the process: define the role of each partner, establish confidentiality of shared information, commit resources, and agree on exception handling and performance measurement.
2. The partners create a joint business plan and establish products to be managed jointly including category role, strategy, and tactics.
3. The manufacturer creates a supply/production plan, based on a single forecast of consumer demand.

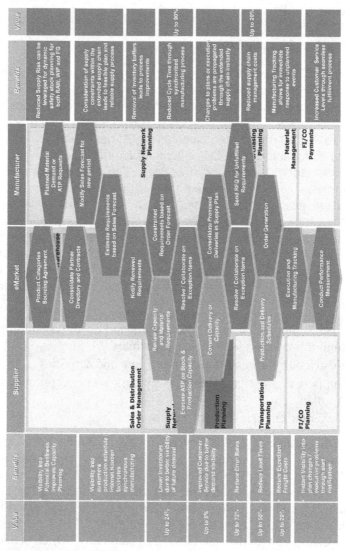

Fig. 11.12. Collaborative Supply Planning Process Scenario with detailed processes, Enabling Applications and Benefits

4. The manufacturer and supplier exchange information on the component requirements and create a joint forecast.
5. The supplier creates a supply/production plan based on the joint forecast.
6. Both partners identify and resolve exceptions particularly those involving the supplier's constraints in delivering specified volumes, creating an interactive loop for revising orders.
7. Both partners continue with succeeding planning steps.

Collaborative Production Execution

Collaborative Production Execution ensures that future material needs are satisfied, resulting in improved order fulfillment. Manufacturers get visibility into suppliers' material availability, schedule and constraints. By calculating dynamic material availability and lead-time using constraints across network, suppliers and subcontractors may optimize their own production schedules resulting in more timely deliveries and minimal delays.

Collaborative Engineering or Product Design

Collaborative Engineering or Product Design, improves the development cycle time for new products and helps develops better products more efficiently. Multinational companies operating globally across multiple time zones characterize the current engineering marketplace. These conglomerates typically outsource many of their standard operations to subcontractors. Collaborative Engineering and Design facilitates the cooperative effort essential for coordinating the engineering and project management tasks of dispersed groups and for involving development partners, contractors, suppliers and customers directly in the product development process. The result is a collaborative environment in which the company responsible collects project-relevant information, publishing it for access by business partners who may or may not be members of an e-Market. The only thing that the partners need on site is a Web browser. All participants in a collaborative scenario receive notification of changes or new project assignments. The collaborative communication is carefully monitored throughout the process, ensuring the right people get the right information at the right time.

11.4.4 Manufacturer-3PL Collaborative Planning Processes

Collaboration among companies and their 3PL providers will focus on joint logistics planning. 3PLs provide transportation shipper services in order to make better use of their transportation equipment and warehousing and distribution center facilities. This might involve collaborative planning to help ensure vehicles are fully loaded by the following:

- Consolidating a shipper's inbound, inter-facility, and outbound shipments
- Combining the shipper's goods with those of another trading partner

These activities involve a shipper electronically sharing the shipment plan with a carrier and comparing it to the availability of equipment, labor, and other transportation resources. Trading partners can support this through joint electronic visibility of transportation resources. Collaboration between a company and 3PLs providing distribution center (DC) services will focus on the productive use of facilities, labor, and equipment. This might involve electronic sharing of DC inventory replenishment plans with analysis to ensure that planned receipts do not overload the receiving function. Plans may also need to be shared to ensure that each DC has enough space to store planned inventories. In addition, 3PL providers can provide insight into the potential for co-sharing of space among trading partners. For example, around the Christmas holidays some of the manufacturer's DCs may be overloaded, providing an opportunity to use a 3PL facility on a temporary basis to correct the problem. This type of collaboration would be further supported by electronic visibility into the availability of DC space and other resources.

Collaborative Transportation Planning

Collaborative Transportation Planning between manufacturers and their carriers allows both partners to streamline their work processes and ultimately benefit from reduced handling costs, greater transparency and greater efficiency. Members of an e-Market may share DC inventory replenishment plans with logistics providers, and inform their carriers about their shipment plans, and the carriers can accept, reject or change shipment requests. Based on current developments within the APS systems sphere, e-Markets are enabling a more full view of the opportunities for transportation by facilitating *Tendering for Bids*[33] and *Advanced Shipping Notifica-*

---

[33] With this function, planners can offer shipments to carriers through the e-Market. A planner can react to the offers made by the carrier and also supervise the status of the tenders. The planner receives tendering statistics and can also judge the service quality of a service agent. The possibility to call for tenders for shipments directly through the e-Market is an additional planning function. The interaction between the planner and the service agent may run completely through the system. Planners can also include carriers who are not

tion[34]. Recently the CPFR model itself has been extended to include Transportation Carriers and 3PLs in a more rounded "3-way" business model. This allows buyer, seller and carrier to come together to exchange key information, provides visibility to status data and conformance to plan, and then provides processes to jointly derive the plan itself.

This new initiative is called *Collaborative Transportation Management*, or CTM. The CTM model can be executed as stand-alone or in parallel with CPFR. CTM progresses through the following activity steps:

- In Step 1, the trading partners establish an agreement to collaborate. This agreement defines the relationship in terms of freight terms (who pays for and controls the carrier relationship) which products will be included, the locations that will be involved, the types of shipments that will be included and the strategies for managing exceptions (this is equal to step 1 of CPFR). This also includes a summary of Key Performance Indicators (KPIs) that will be used to measure the relationship to ensure that satisfaction is being achieved all-around, which may or may not align with CPFR.

---

members of the current e-Hub in the decision-making process - the carriers process the data through their own systems. System access is monitored using user safety profiles and authorization objects that are assigned specially for the tender status tasks. The service agents can then call up only those shipments that were offered personally to them.

[34] Vendors can use inbound-delivery processing through the Internet to create and process shipping notifications for the customer. The system ensures that a vendor can only select purchase orders that belong to him/her. The user can create and change shipping notifications, which are reflected in the customer's system as inbound deliveries and contain basic data such as the delivery date or delivery quantity of these inbound deliveries. At the point of shipping notification entry, a list appears to the vendor that displays all purchase orders and scheduling agreements that are relevant to that vendor. After the delivery date and the unique identification number have been entered, an inbound delivery for the customer is generated. The customer and the vendor can also modify these deliveries at a later stage, and all parties can view any changes in real time. This process is an alternative to the previous order notification method through EDI and produces the same result.

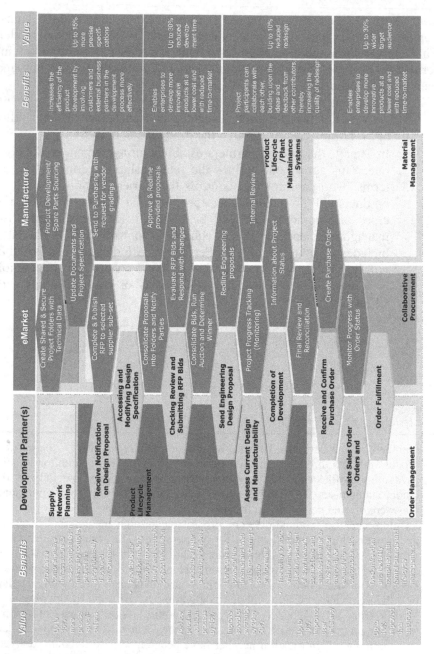

Fig. 11.13. Collaborative Engineering, Tendering and Project Management –
Processes, Enabling Applications and Benefits

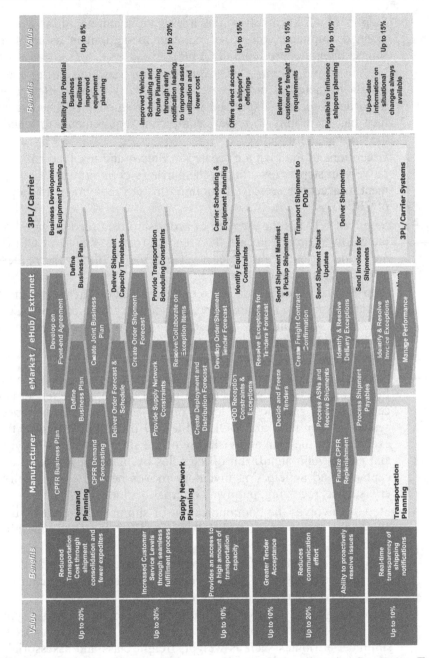

Fig. 11.14. CPFR-related Collaborative Transportation Planning Processes, Enabling Applications and Benefits

- Step 2 of CTM involves the aggregate planning phase where planned shipment volume is matched to equipment asset plans (this is integrated to step 5 of CPFR).
- Next in Step 3, *Create Order Shipment Forecast*, the carrier gains insight to increases or decreases in planned volumes reflected in the order forecast - expressed in terms of shipments. The carrier then has the ability to review equipment requirements to handle the shipments forecasted.
- Exceptions to the plan are created in Step 4 and resolved collaboratively in step 5, resulting in the carrier's commitment of equipment to accept the resolved volume (this is synonymous to Step 8 of CPFR).
- Step 6 of CTM is the creation of order/shipment tenders based on the resolved order forecast. The tenders (part of step 9 of CPFR) are made earlier in the process, in order to facilitate the highest level of acceptance by the carrier.
- Step 7 is the identification of exceptions based on latest equipment availability, pickup requirements and delivery requirements. Collaboration will eliminate unnecessary wait time and subsequent charges, and will improve overall efficiencies.
- Step 8 is the acceptance of a tender.
- Step 9 is creation of the final shipment contracts for specific freight orders. This signifies the results of collaborative tender acceptance and specifies the terms of the agreement, today represented as manifests and BOL's. Steps 9 through 11 involve the execution of the plan and visibility of the shipment status. Buyers and sellers gain significant efficiencies by planning shipment acceptance and anticipating inventory moves beyond carrier delivery, such as receiving, put-away or cross docking.
- Step 10 involves the communication of shipment attributes (such as weight, line items, freight classes and assessorials) and shipment status.
- In Step 11, shipment status is continually updated as to progress and projected delivery, creating delivery exceptions and changes to be resolved interactively between the parties.
- Steps 12 and 13 involve the traditional payment process. Typically, there are differences between carriers and shippers as to shipment attributes such as weight, freight class and assessorials. The information for these exceptions is provided in Step 10 and collaboratively resolved through messaging in steps 12 and 13.
- Finally, the partners' measure and report performance in Step 14 against KPIs included in the CTM agreement from Step 1.

With this process we finalized our guided tour of some central collaborative planning processes that may be executed between various partners through an e-Market. In the next section, we will present what are the tangible benefits achieved by implementing collaborative planning. CPFR has been a major instigator of collaboration pilot projects around the world.

11.5 Benefits of Collaboration in e-Supply Networks

While popular literature and business press has touted the benefits of collaborative planning, recent pilots have shown that the actual benefits far exceed expectations. Manufacturers that implemented collaborative planning achieved: reduced inventory levels (18-40%), increased inventory turns (20-70%), reduced production cycle times (up 67% reduction), reduced returns (5-20%), improved forecast accuracy (7-20%), reduced freight cost (18-20%), and lower overall distribution costs (10-30%).

On the other hand, downstream partners achieved: Increased sales (12-40%), increased buyer productivity (40%), improved customer service levels (up to 22%) and in-stock availability (as much as 8%), reduced overall inventory levels (18-40%), and increased inventory turns (10-30%). These are significant results reached only for subsets of various product categories traded between partners. Another still not fully understood benefit is that improvement of partner communications, release enormous amounts of time, which partners can spend improving customer relationships and handling exceptions.

CPFR is fast becoming the most explored model of downstream collaborative planning. In comparison with Vendor Managed Inventory or other initiatives that has gone before, pilot implementations of collaborative processes with the CPFR methodology (in Wall-Mart, Kimberly Clarke, HP, P&G, Nabisco and others – see CPFR Roadmap, 2000) have shown significant benefits, to both buyer (retailer, manufacturer etc.) and seller (manufacturer, suppler etc)[35].

---

[35] A survey by Industry Directions (April 2000), found that over two-thirds of those surveyed (130 Fortune-500 corporations) are actively involved in CPFR activities or pilot research. About one-quarter of the respondents have a CPFR pilot underway or plan to start a pilot within the next 6 months.

Table 11.4. Summary of benefits from the various CPFR pilots

Value Lever	Operational / Financial Impact	Benefit to Buyer	Benefit to Seller	Driver / Enabler
Collaborative Planning	Clearly defined performance metrics			Define roles & responsibilities for each partner
	Agreed joint category strategies			Develop joint business plans
Collaborative Forecasting	Improved Forecast Accuracy And Timeliness	Up to 20%		Increase forecast accuracy via shared downstream/ upstream information
	Improved Supply Visibility			Improve supply information
	Improved Demand Visibility			Improve demand information
	Improved Exception-Handling			Enhance Communication between trading partners
Collaborative Replenishment	Reduced Lead Times Through "Pull" Replenishment		50% reduction	Increase downstream demand visibility
	Higher In-Stock Availability	5-8%		Reduce order cycle times, Improve in-stock position
	Reduced Production Cycle Times		Up 67% reduction	Improve procurement co-ordination, Supply contracts for new products
	Reduced Transaction Costs	50-75%	50-75%	Flexible aggregate planning
	Reduced Inventory Costs		13%	Increase pipeline visibility to eliminate buffer inventory
	Lower Overall Inventory Levels	10-30% reduction	18-40% reduction	Improve match of supply w/ demand
	Increased Inventory Turns	10-30%	20-70%	Improve sell through and cycle times
	Reduction In Returns		5-20%	Improve downstream demand visibility
	Decreased Obsolescence Rates		5-10%	Improve downstream demand visibility
	Reduced Transportation Costs		2-10%	Improve fulfilment and procurement co-ordination
	Improved Replenishment Cycles			Improve manufacturing planning and efficiencies
	Improved Customer Service Levels	10-30%		Improve demand information
	Improved Reliability Of Supply			Improve procurement co-ordination
	Increased Sales	20-70%		Improve order fill rates via pipeline visibility & reduce lead times
	Reduced Lost Sales			Improve demand information
	Improved Order Fulfilment			Improve fulfilment co-ordination

11.6 Implementing Collaborative Planning

In order to implement Collaborative Planning, companies have to realize that they are part of a broader business network or ecosystem, which is per definition collaborative. Beyond an in-depth understanding of their core competences, members of such communities have to standardize their information and process infrastructures. Most companies need to overcome a number of barriers in order to successfully implement or participate in such Collaborative e-Markets: Variability[36], Scalability[37], Uncertainty[38] and

---

[36] Since there is no single business process that fits CPFR for all consumer goods and retail firms, there is a set of CPFR alternatives that needs to be mixed and matched by trading partners to fit their needs.

[37] Most early pilots have been managed with sheer labor. According to Wegmans' D'Arezzo: "We achieved CPFR with Nabisco using paper and pencil and hard work. To do with a lot of suppliers in a lot of product categories, we need technology to automate it."

Change Management[39]. Finally, collaboration requires some semantic synchronization (e.g. Master data, units of measure). Business partners have to agree on standards to be used for routine collaborative processes - *who is doing what, when and where, plus who is responsible?*

11.6.1 Organizational Implementation Considerations

CPFR pilots have identified a number of recurring challenges that have to be resolved in order to make collaboration a success. These are a) *Mutual trust,* b) *Sharing of savings and risks,* c) *Common performance metrics,* d) *Adoption of inter-enterprise business processes,* and e) Striving to reach *critical mass* ASAP. Case companies that have already implemented the CPFR business model, have identified some critical cross-functional issues, that need to be understood and addressed by potential partners pursuing closer collaborative arrangements:

- Building *trust* and *collaboration* among trading partners
- Reducing channel *conflict* (by mapping and handling potential exceptions)
- Enhancing channel services
- Pricing based on market conditions and value versus standard pricing
- Responding to customer needs and demands versus the pushing of products from the supply chain to customers.
- Adopting standard business documents, terms, and processes.

Collaborative e-Markets help companies do more work with fewer people by automating routine communication and offloading simple services to customers. Collaborative processes transform organizations in that they

---

[38] Many trading and consumer goods firms are naturally reluctant to share the plans in advance, fearing that they will somehow fall into the hands of competitors. While CPFR eliminates significant uncertainties and inventories across the entire supply/demand chain, but a preliminary top-level agreement on how savings are shared is critical. According to Jim Uchneat: "Vision does not create success as much as writing down, such as who gets what dollars of savings".

[39] The real key to successful CPFR implementation is forging cultural alliances rather than traditional adverse relationships. Benchmarking Partners: "A company must itself have an culture of openness, and it must have a leader at an upper level who makes a jump over the hurdle of "we can't let them have that information."

allow customers and suppliers to serve themselves[40], reach new customers without adding staff[41], and automate common business processes[42].

11.6.2 Technological Implementation Considerations

Deployment of collaboration in an e-Markets environment offers tremendous benefits, as described above. At the same time, the increased level of aggregation that *Collaborative e-Markets* provide results in challenges for the traditional hub-and-spoke-based collaboration planning processes. Some of these challenges are described below:

1. **Data normalization** - Exchange participants need to agree on:
   Conventions for representing time buckets. The greatest challenge of a Collaborative Planning in an e-Market is managing the diversity of time buckets used by different trading partners[43].
   Product codes. Capability to utilize Global Trading Item Numbers

---

[40] When customers can use services provided by collaborative processes to embed your product configurator or order status service inside their operations, they can answer questions on their own time rather than waiting for you to help. Customers get faster, cheaper responses, and you get to keep call center staff to a minimum. Moreover, such interactions allow companies to get feedback on what customers need ("Organic Business" by Ted Schadle, Charles Rutstein, Carey E. Schwaber, Forrster Research, April 2004).

[41] Since e-Market services use standards to project a company's data down the wire, they can reach new customers at minimal marginal cost. ("Organic Business" by Ted Schadle, Charles Rutstein, Carey E. Schwaber, Forrester Research, April 2004).

[42] Firms are stuck maintaining — and training every new employee on — scads of standalone applications. The swivel-chair integration required to tap all those systems will disappear as firms like Pfizer replace redundant manual approval processes with a single approval engine. The results? Faster training, higher productivity, and more lights-out operations ("Organic Business" by Ted Schadle, Charles Rutstein, Carey E. Schwaber, Forrester Research, April 2004).

[43] Here there are many alternatives: A) Standardize on common weekly and monthly calendars on the exchange (for example, normalizing week boundaries on the exchange to Sunday to Saturday). Weekly and monthly data is mapped from enterprise-specific calendars as it is sent or received. B) Disaggregate weekly and monthly data to daily buckets as it is received, using allocation rules. The CPFR solution then provides enterprise-specific calendar views of this data on the exchange. C) Bring data onto the exchange in as-is form. Use allocation rules to present aggregate views, and allow exception messages to diverge between the buyer and the seller.

(GTIN) that incorporate U.P.C. codes, EAN codes, and new identifiers. (See the Global Commerce Initiative guidelines)

Location codes. Ability to support Global Location Numbers (GLN) instead of DUNS.

Scorecarding criteria that will be used to rank trading partners.

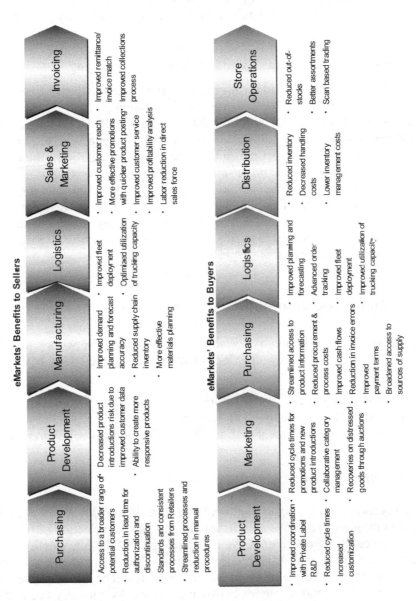

Fig. 11.15. Summary of e-Market Collaboration Benefits

2. **Internationalization** - Ability to simultaneously handle different: Languages, Currencies, Date formats, and Time zones.
3. **Interfaces** with Global Commerce Initiative (GCI) approved processes for item alignment, party alignment, and purchase order.
4. **Security** at the enterprise, user and exchange layer. The security model and management of the exchange must meet the highest security standards because data is represented from many buyers and sellers.
5. **Interoperability** - CPFR XML schema support to guarantee consistent hub-to-hub and advanced peer-to-hub-to-peer messaging.

Companies have to bridge information from their internal Enterprise Resource Planning (ERP), Advanced Planning and Scheduling (APS/SCM), Customer Relationship Management (CRM), and legacy applications to one or many e-Markets, either initially via a Web browser, or eventually via system-to-system integration. Thus, the building blocks of such integration is a) *common business documents and transactions*, and b) *common semantics, taxonomies and standards* (both data and process). This enables multiple IT applications to work in unison to enable end-to-end processes triggered either internally or externally. Such IT-enabled process orientation is a prerequisite for enabling collaboration across internal departments and external partners.

11.7 Some Implementation Considerations

The benefits of Collaborative Planning in e-Supply Networks may be clear, but there are 3 key interdependent questions that need to be addressed:

1. Who should you collaborate with?
2. How should you go about collaboration?
3. What are the requirements for and the implications of collaboration?

11.7.1 Who should you collaborate with?

Collaboration requires significant investment in time and resource for both partners in order to achieve significant benefits, so the selection of partners should be carefully considered. New technology and the introduction of e-Markets and ASPs may have reduced some of the technical risk and cost, but for collaborative planning to be effective it needs the alignment of people, processes and resources between partners. As with any other critical business decision the cost and benefits should be carefully assessed, and in this case this is for at least 2 partners. The result is that collaboration should be targeted at your long term trading partners, for key products, where the product and service are primary buying factors.

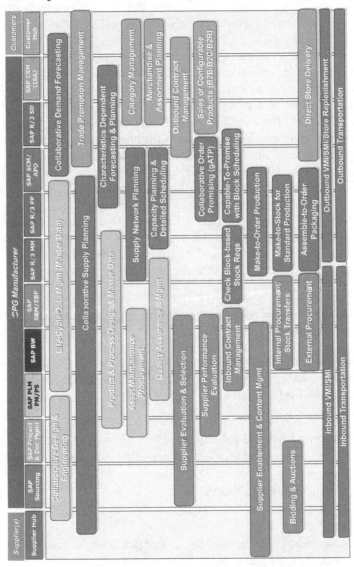

Fig. 11.16. End-to-end processes and application enablement (this example with SAP applications)

11.7.2 How should you go about collaboration?

There are several possible strategies for collaboration. Historically implementation of CPFR can be seen as partnership, process or technology lead. Close trading partners have recognized the mutual benefits of collaborative planning, and have evolved the processes for this. This has had the benefit of building on the trust and working knowledge, which are key to success; however, the processes and any technical solutions may be inefficient and not easily transferable to other partners. Alternatively, companies have designed CPFR processes into their ways of working, and rolled these out to their key partners, for example, motor manufacturers and their suppliers. However, this may be dependent on a dominant player, and be less than fully collaborative. Latterly Advanced Planning Systems (APS), and the Internet have provided the tools that have driven many CPFR implementations. The ideal strategy combines the right balance of partnership, appropriate process design and use of enabling technology.

One feature of many successful CPFR implementations, and inherent in a partnership lead approach, is the use of pilots. This allows the evolution of the right process, and the understanding of the changes needed to organization, roles, and performance measures, as well as technology.

11.7.3 What are the requirements for and implications of implementation?

Case companies that have already implemented the CPFR business model, have identified some critical cross-functional issues, that need to be understood and addressed by potential partners pursuing closer collaborative arrangements:

- Building trust and collaboration among trading partners
- Reducing channel conflict (by mapping and handling potential exceptions)
- Enhancing channel services
- Pricing based on market conditions and value, versus standard pricing
- Responding to customer needs and demands, versus the pushing of products from the supply chain to customers.
- Adopting standard business documents, terms, and processes.

11.8 Epilogue

Enablement of complex, collaborative e-Supply networks with information infrastructures may lead to a highly-probable, though slow, Lamarckian evolutionary process, rather than a revolutionary inflexion point of current business practices among Industrial Organizations. Notwithstanding, the value of using the Internet-based IT - or any supply chain/relationship management tool - may not significantly improve until the company re-invents itself to embrace internal and external transformation – not an easy undertaking!

e-Business is changing the Industrial Age models of customer acquisition, procurement, pricing, and customer satisfaction as well as how we measure the performance of a corporation. Focus on the customer is all consuming; customers want to buy products anytime, anywhere, cheap and fast, and fulfillment processes must be structured to meet these demanding requirements. Companies are simply recognizing that the old rules will not give them the continued success that they had enjoyed, but instead, new ways and protocols are emerging.

What we have presented so far is the substantial structural changes that are underway within the area of e-Supply Networks and their catalysts, e-Markets. The functionally driven silos present in many contemporary supply chains are being transformed, and replaced by more streamlined, electronically based processes. Internet and associated technologies such as XML have revolutionized inter-enterprise business processes by enabling seamless information exchange between business partners. High volumes of data can be transferred at low cost, and even minor business partners can exchange information in an economic manner. Interactive on-line access to each others' systems can be achieved easily via a conventional Internet browser. Thus, propelled by the accelerating permeation of information and communication technologies into intra-organizational processes that also enable inter-organizational collaboration, companies are clustering into private and collaborative marketplaces to conduct their business. These Collaborative e-Markets are entering in between the buyer-seller relationship and are bound to change the rules of the competitive game. In short, we believe that the development, promotion, and adoption of these network and business models, will maximize the impact of e-Business in most industries, and enable companies and customers to begin reaping the benefits of the new digital economy, where most activities will happen On Demand!

References

Amor, D. (1999) *The E-business (R)evolution*, Prentice Hall

Chisholm, R.F. (1998) *Developing Network Organizations: Learning from Practice and Theory*, Addison Wesley Longman

Christopher, M. (1994) *New directions in Logistics, Logistics and Distribution Planning – strategies for management*; edited by James Cooper, Kogan Page Limited, 2nd Edition

CPFR Guidelines (1998) *CPFR Collaborative Planning, Forecasting, and Replenishment*, VICS – Voluntary Interindustry Commerce Standards, http://www. cpfr.org

CPFR Roadmap (1999) *Roadmap to CPFR: The Case Studies*, VICS – Voluntary Inter-Industry Commerce Standards, http://www.cpfr.org

Curran, T.A., Ladd, A. (1999) *SAP R/3 Business Blueprint: Understanding Enterprise Supply Chain Management*, Prentice Hall

Ciborra, C. (2002) *The Labyrinths of Information – Challenging the Wisdom of systems*, Oxford: Oxford University Press

Corsten, D. and Hofstetter, J.S. (2001) Supplier Relationship Management, in: Belz, C. Mühlmeyer, J., (eds.), *Key Supplier Management*, Thexis: St. Gallen

Dellarocas C., Klein M. (2000) A Knowledge-Based Approach for Handling Exceptions in Business Processes, Working paper, Center for Coordination Science, Sloan School of Management, Massachusetts Institute of Technology

Easton, G., Araujo, L. (2003) Evaluating the impact of B2B e-commerce – a contingent approach, *Industrial Marketing Management*, Volume 32, Issue 5, July 2003, p. 431–439.

Egelhoff, W. G. (1991) Information-processing theory and the Multinational enterprise, *Journal of International Business Studies*, Vol 4

Ford, D. (ed.) (1997) *Understanding Business Markets*, 2nd edition, Dryden Press

Ford, D. (ed.) (2002) *Understanding Business Marketing and Purchasing*, 3rd edition, Thomson Learning

Ford, D., Gadde L., Håkansson H., Lundgren A. (1998) *Managing Business Relationships*, John Wiley

Ford, D. and Håkansson, H. (1999) How should companies interact?, Conference Paper, IMP 1999, Dublin.

Gadde, L. and Håkansson, H., (2000) Supply Network Strategies, Conference Paper, IMP 1999, Dublin.

Gadde, L. and Håkansson, H., (2001) *Supply Network Strategies*, John Wiley

Gadde. L. E., Håkansson, H., Jahre, M., Persson, G. (2002) More instead of less: Strategies for use of logistics resources, *Journal on Chain and Network Science* 2: 81–91

Ghoshal, S., Gratton L. (2002) Integrating the Enterprise, *Sloan Management Review*, Fall 2002

Goede, R., De Villiers, C. (2003) The applicability of grounded theory as research methodology in studies on the use of methodologies in IS practices, *Proceedings of SAICSIT 2003*, 208 –217

Goldman-Sachs (1999) *B2B:2B or Not 2B?*, Goldman Sachs Investment Research

Hanseth, O. (1996) Information Technology as Infrastructure, Ph.D. Thesis, University of Gothenburg

Harland, C.M. (1996) Supply Chain Management: Relationships, Chains and Networks, *British Journal of Management* 7: 63–S80 (Special Issue)

Hartmann, E., Ritter T. and Gemuenden, G.E. (2002) The Fit between Purchase Situation and B2B E-Marketplaces and its Impact on Relationship Success, Conference Paper, IMP Annual Conference 2002

Hedaa, L., and Törnroos, J. (1997) Understanding Event-based Business Networks, MPP Working Paper, 1997

Hedaa, L. (1998) Atoms of Interaction – Demigraphic analysis of business networks, MPP Working Paper, 1998

Hedberg, B., Dahlgren, G., Hansson, J., Olve, N. (eds.) (1997) *Virtual Organizations and Beyond: Discover Imaginary Systems*, Series in Practical Strategy, John Wiley

Henderson, J.C., Venkatraman, N., Oldach, S. (1996) Aligning Business and IT strategies, in: Luftman, J., *Competing in the information age: strategic alignment in practice*, Oxford University Press, New York

Huxham, C. (ed.) (1996) *Creating Collaborative Advantage*, Sage Publishers, London

Håkansson, H., Snehota, I. (1995) *Developing Relationships in Business Networks*, International Thomson Business Press, London

Kalakota, R., Robinson, M. (1999) *E-Business: Roadmap for Success*, Addison-Wesley Information Technology Series, Addison-Wesley

Klein, S. (1996) The Configuration of Inter-organizational Relations, *European Journal of Information Systems* 5, pp. 92–102

Lawrence, P.R., Lorsch, J.W. (1967) *Organization and Environment: Managing Differentiation and Integration*, Homewood

Magretta, J. (1998) The power of virtual integration: an interview with Dell Computer's Michael Dell, *Harvard Business Review* 76: 72–84

Malone, T. (1987) Modeling Coordination in Organizations and Markets, *Management Science*, 33, 1317–1332.

Malone, T. W., Crowston, K., Lee, J., Pentland, B. (1993*) Tools for inventing organizations: Toward a handbook of organizational processes, Proceedings of the 2nd IEEE Workshop on Enabling Technologies Infrastructure for Collaborative Enterprises,* Morgantown, WV

Malone, T.W., Crowston, K. (1994) The Interdisciplinary Study of Coordination, *ACM Computing Surveys*, 26 (1): 87–119

Nohria, N. and Eccles, R. G. (1992) *Network and Organizations*, Harvard University Press, Boston, MA

Nøkkentved, C. (2003) Implementing Supplier Relationship Management: A Conceptual Framework Of Value Creation In Collaborative Supply Networks, IBM BCS Institute of Business Value, Research Note

Nøkkentved, C., Hedaa, L. (2001) Collaborative Planning Processes in e-Markets, Conference Paper, IMP Annual Conference 2001.

Nøkkentved, C., Hedaa, L. (2000) Collaborative Processes in e-Supply Networks, Conference Paper, IMP Annual Conference 2000.

Poirier, C. (1999) *Advanced Supply Chain Management : How to Build a Sustained Competition*, Publishers' Group West

SAP APO (1998) Advanced Planner & Optimizer, Product Whitepaper, SAP AG, Walldorf, Gemany

SCOR (1998) Supply-Chain Operations Reference-Model, Version 3, Supply-Chain Council, Pittsburgh, PA, September 1998.

Scheer, A.-W. (1998) *ARIS – Business Process Frameworks*, Springer Verlag, Berlin

Scully, A., Woods, W. (1999) B2B Exchanges: The Killer Application in the Business-to-Business Internet Revolution, ISI 1999.

Simon, H.A. (1997) *Administrative Behavior*, Free Press, 4th Edition, (1st Edition: 1945), New York

Shapiro, C., Varian, H.R. (1998) *Information Rules: A Strategic Guide to the Network Economy*, Harvard Business School Press, Boston, MA

Simchi-Levi, D., Kaminsky, P., Simchi-Levi, E. (1999) *Designing and Managing the Supply Chain: Concepts, Strategies and Case Studies*, Irwin/McGraw-Hill

Snow, C. C., Miles, R. E., and Coleman, H. J. (1992) Managing 21st Century Network Organizations. *Organizational Dynamics*, 20(3): 5–20.

Stewart, G. (1997) Supply-chain operations reference model (SCOR*)*, *Logistics Information Management* 10(2): 62–67

Tapscott, D. (1995) *The Digital Economy*, McGraw-Hill, New York, NY

Tapscott, D., Ticoll, D.,Lowy, A. (2000) *Digital Capital: Harnessing the power of business webs*, NB Publishing, London

Weill, P., Broadbent, M. (1998) *Leveraging the New Infrastructure*, Harvard Business School Press, Boston, MA

Österle, H., Fleisch, E., Alt, R. (1999) *Business Networking – Shaping Enterprise Relationships on the Internet*, Springer Verlag

12 Sense and Respond Business Performance Management

Steve Buckley, Markus Ettl, Grace Lin and Ko-Yang Wang

12.1 Introduction

Today's market and business environments are inherently complex, dynamic and global. Customers are becoming more informed and demanding. To stay competitive, enterprises must improve their flexibility, efficiency and responsiveness by transforming their business and operational models.

During the past 20 years, supply chain management has evolved from the internal efficiency improvement and cost cutting focus of the 80's, to the limited information sharing of the extended supply chains and the ERP implementation for transaction efficiency of the 90's. However, a major issue with the ERP system is its lack of flexibility and speed to support decision making throughout the internal and extended supply chain to meet changing business requirements. By the mid 90's, Advanced Planning and Scheduling (APS) tools, implemented with legacy and ERP systems, allowed "what-if" analysis and optimization of supply chains during planning and execution cycles. In the late 90's, the development of packaged applications, e-Commerce and e-Business offered Internet connectivity and limited capability for supply chain collaboration and near real-time information sharing and decision making.

However, despite the implementation of the supply chain management tools and the Internet connectivity, enterprises still found that they often sub-optimize their operations. Furthermore, the ROI of supply chain management package implementations is constantly being questioned. Forrester reported (based on interviews of 25 firms) that companies overspent on supply chain optimization packages and got diminished returns: 80% of the companies spent more time than expected; and, on the average, companies spent 74% over budget to implement supply chain optimization tools (Lawrie 2003).

In fast changing business environments, business and technical problems can occur anytime, and at every level. Lack of information visibility across internal and external supply chains, insufficient customer collaboration, inability to leverage knowledge and manage uncertainty, overly rigid business processes, and lack of infrastructure flexibility can all cause major business disruptions and inefficiencies. Local supply chain optimization based on incomplete information under rigid top-down planning models can not only result in sub-optimization, but can also cause adverse effects.

The key to a successful adaptive organization is ensuring continued focus on responsiveness and agility. New business models enabled by real-time business process management are evolving. They present new opportunities that enterprises hope to embrace to enhance their competitiveness. At the same time, converging social and technological trends are changing the nature of decision-making. The Internet has caused an explosion of information availability. Pervasive computing and wireless technology have added to the information pile by drawing from formerly isolated sources of data. Improvements in network bandwidth and processor technology have reduced information latencies, and enabled businesses to place large numbers of sensors. Flexible interconnect technologies such as Web Services have made it easier for one network to pass information onto another network. However, an abundance of information, and the ability to respond rapidly to events do not guarantee success.

The successful supply chain optimization of enterprises also depends on their ability to streamline operations, while being able to process information intelligently and holistically, so they can respond proactively and effectively. This includes fully understanding the needs of customers and business partners, and the capabilities of employees, as well as analyzing risks and opportunities in a changing environment.

After the IBM Extended Supply Chain R&D group successfully helped IBM Personal System Group and other divisions transform their supply chains from Make-to-Plan to Configure-to-Order (Lin et al. 2000), the group started to explore more flexible and responsive models. In doing so, they included lessons learned from a number of supply chain transformation efforts both within IBM internal groups and their partners. Sense and Respond Business Performance Management was initiated in 1999 at IBM Research to build an open and adaptive framework, using intelligent decision making and IT technology for business optimization (Lin et al. 2002; Lin et al. 2004). This effort departed from the traditional extended supply chain models. Traditional extended supply chain management systems are based on static, structured information within a rigid planning cycle, while Sense and Respond Business Performance Management orchestrates dynamic, structured and unstructured information within a continuous, adaptive event-based planning process. Traditional extended supply chain management focuses on supply chain planning and execution. Sense and Respond Business Performance Management not only performs supply chain planning, but also determines business rules and policies and orchestrates among the value partners to achieve better overall performance. Traditional value chain management responds to environmental changes reactively, while the Sense and Respond framework utilizes a real-time, predictive and proactive modeling capability to address potential issues.

The rest of the chapter is organized as follows. We start with an overview of related work in section 12.2. The four key requirements for an enterprise to become more adaptive are discussed in section 12.3. In section 12.4, we introduce two core aspects of the Sense and Respond Business Performance Management framework: a model-driven capability design and an architectural framework of loosely coupled components for adaptive business management. In section 12.5, we describe two pilot implementations where the Sense and Respond Business Performance Management framework was applied. In the pilot with IBM's Personal Computing Division, we describe how to use order trends to provide early warnings of constraints and excesses of PC components; in the second pilot with IBM's Microelectronics Division, we describe how to use an end-to-end supply chain model to support event-driven management of inventory and customer order fulfillment. We conclude the chapter with a summary in section 12.6.

12.2 Related Work

The first reference to the term "Sense and Respond Organization" we found was in the book entitled "Sense and Respond: Capturing Value in the Network Era" (Bradley et al. 1998). This book focuses on two components of value creation: electronically sensing customers' needs in real time and using the electronic connection and shared infrastructures to respond to those needs.

In his 1999 book entitled, "Adaptive Enterprise: Creating and Leading Sense-and-Respond Organizations", Stephan Haeckel of IBM defined the Sense-and-Respond business organizational change model (Haeckel 1999). He describes the transformation from a Make-and-Sell organization to a Sense-and-Respond organization and advocates a new form of strategic transformation based on roles and responsibilities. In his view, organizational hierarchy is replaced by a dynamically configured network of modular capabilities. Governance is performed on the basis of context and coordination by people in roles accountable for outcomes rather than by command and control.

AMR Research defines Supply Chain Event Management, or SCEM, as a class of supply-chain-management software that allows companies to respond to unplanned events on an exception basis (Bittner 2000; Suleski and Quirk 2001). SCEM comprises integrated software functionality supporting the five business processes Monitor, Notify, Simulate, Control and Measure.

Business Activity Monitoring, or BAM, defined in 2002 by Gartner Group is a class of software that provides real-time access to critical business performance indicators to improve the speed and effectiveness of business operations (Correia and Schroder 2002; Lehmann 2003). It focuses on the IT aspect of the adaptive enterprise.

Lee and Amaral (2002) describe Supply Chain Performance Management (SCPM) as a cycle consisting of identifying supply chain problem areas, understanding root causes, responding to problems with corrective actions, and continuously validating data, processes and actions. The authors describe how the approach was used by two large electronics manufacturers to improve the velocity of their extended supply chain.

Sense and Respond Business Performance Management addresses the full spectrum of SCEM and takes it several steps further. It enhances the global visibility by focusing on the collaborative interactions based on the degree of trust on data, people, and organization. The model integrates supply chain planning with a dynamic sense-and-respond control model, utilizing an agent-based framework that supports different business and execution models. It allows enterprises to adaptively use the most effective model to address their value chain needs. It enhances the event-based management by marrying real time decision support with end-to-end performance and risk management. Further, it helps realize operational business designs through business process integration, automation technology and Web services, and helps partners to integrate their processes and optimize supply chain collaboration and enables intelligent decision-making on events and prediction of future value chain performance.

12.3 Sense and Respond Business Performance Management Roadmap

An adaptive business senses changes in the environment and in the needs of customers, employees and business partners. Adaptive businesses focus on core competencies and support an open and integrated operating environment to collaborate with customers and suppliers. Being adaptive means being able to sense changing business conditions and customers' needs and respond with speed and intelligence.

The adaptive business approach provides promising solutions to many of the challenges that companies face throughout the value chains. As in any major business transformation, becoming an adaptive e-business requires establishing a strategy and a roadmap (IBM 2003). To become adaptive, enterprises need the following capabilities:

- *Automation.* Proactively developing a better understanding of transactional data representing customers' needs while also monitoring environmental factors.
- *Visibility.* Integrating data and applications within the enterprise as well as with business partners, suppliers and customers to increase visibility and operational efficiency.
- *Control.* Creating an IT infrastructure that fully supports business goals and has the intelligence to help transform ways to do business so that enterprises can react with agility to the changing environment.
- *Adaptiveness.* Developing competitive advantages through adaptive optimization supported by dynamic tradeoff analysis and cross functional collaborations.

Sense and Respond Business Performance Management addresses the full spectrum of all of the above concepts. It integrates value chain planning with a dynamic Sense-and-Respond control model, utilizing an agent-based framework that supports different business and execution models. It allows enterprises to adaptively use the most effective model to address their value chain needs. Further, it helps realize operational business designs through business process integration, automation technology, and Web Services, and helps partners to integrate their processes and optimize supply chain collaboration.

12.3.1 Automation

A key requirement to an optimized execution of the extended value chain is the ability to collect, maintain and manage information linked to business partners, customers and suppliers. Information such as customer profiles, supplier status, customer demand, product information, planning data, current inventory, capacity, pricing, and product and process cost must be accurate and timely. Transactional data comes from business applications and is usually process-related. Derived data may be obtained from data services, partners, internal sensors and post-analysis of historical data. Intelligent decision-making is only possible when up-to-date and accurate information is available.

In today's business environment, data is often stored in a variety of formats using various tools and may be quite fragmented. Data integrity problems come from different sources: fragmentation, incompleteness, data unavailability, data latency (delays in data arrival), nonstandard data models and lack of trust in sharing data among business partners. Techniques to ensure good data quality include data cleansing, measurements, information integration and analytic processing.

The Sense and Respond Business Performance Management architecture leverages J2EE-compliant application servers to acquire data, analyze data to detect events, and invoke actions in response to events. It enables monitoring of data within and outside the enterprise such as data from databases, JMS message queues, ftp repositories, and web services. Advanced analytics leverage the transactional data to predict critical events and invoke responses when such events occur, which ultimately enables the development of contingency plans before events impact the value chain.

12.3.2 Visibility

Visibility across the enterprise requires real-time information, rationalization, aggregation, performance analysis capability, workflow technology and dashboard technology. Limited visibility can prevent organizations from optimizing their value chain or internal operations, and may be a direct cause of excess inventory. It can lead to bad decisions with costly results. On the other hand, excessive information from data warehouses, automatic sensors, partners, portals, etc. can be overwhelming and make it difficult to identify the important data from the mundane data.

Companies that can understand their data needs, proactively collect useful data, analyze and manage data utilizing filtering and data aggregation techniques, and use data to derive intelligent information will have significant competitive advantage.

Dashboards and portals help to aggregate and synchronize enterprise information and enable workflow-based information display for users based on their roles. They support the presentation of performance and supporting information in standard formats for aggregation and analysis. Transactional visibility and dashboard displays of information can be enhanced with well-defined business logic. Alerts can be generated for timely decision support. Most importantly, workflows with exception-based business logic and rules can identify where critical points in the supply chain processes may require immediate intervention (e.g. potentially late or missed shipment, supplier quantities received not equal to ordered quantities). All too often when individual value chain components are measured independently, functional performance meets or exceeds thresholds but in the aggregate, the entire supply chain remains sub-optimal due to lack of visibility, synchronization and control.

12.3.3 Control

A control is a closed-loop feedback mechanism that drives a business process towards performance goals. A control utilizes available data, business logic, and analytics to determine whether actions are required in response to disturbances in the business environment and recommend actions. A significant fluctuation in demand, for example may initiate a response that the production schedule be re-optimized and then send the appropriate information to the appropriate party. That same fluctuation may also send collaborative messages and even correcting transactions to trading partners, such as logistics service providers and suppliers. An example might be a shipment request, or a purchase order change.

Potential control actions can be simulated based on historical responses, business rules, business analytics, what-if-analysis, risk analysis or predictive modeling. The effectiveness of a control depends upon the latency and accuracy of available data, the completeness of the data, and the sophistication of the business logic and analytics. Because the environment is constantly changing, a control must allow its business logic to be updated dynamically without programming interruption.

To achieve optimal business performance in a dynamically changing environment, enterprises need to exercise control at all levels (strategic, tactical and operational) to determine the best short-term and long-term course for their value chain. Adaptiveness implies that the organization and its processes are "adaptive", responsive and agile. Multiple functional controls and cross-functional demand and supply signals are integrated at all levels within the organization to collaboratively optimize strategy, business policies and operational decisions.

12.3.4 Adaptiveness

Adaptiveness involves the use of predictive modeling and learning based on historical performance coupled with real-time information. Adaptiveness requires two feedback loops:

-   A robust version of the basic control loop described in the previous section
-   A loop that detects inaccuracies and weaknesses in the control model and adjusts the control model accordingly. Control model

adjustments may be cascaded to strategy models and business process models.

Adaptiveness involves creative coupling of decision-making technology such as analytical models, simulation, and pattern recognition, data mining and learning algorithms. Visibility to relevant value chain data is critical. Advanced mining and forecasting techniques enable enterprises to sense trends for longer-term capability networks planning and events for short-term response optimization. Analytical techniques for strategic sourcing, inventory management, dynamic pricing and risk management enable intelligent decision-making and predictions of future supply chain performance. This enables enterprises to focus on core competencies and support an open and integrated operating environment to collaborate with customers and suppliers.

12.4 Sense and Respond Business Performance Management Framework

Since the mid-1990's, IBM has been developing and refining software architectures and tools to support Sense-and-Respond business management. In this section we describe model-driven capability design, a core aspect of our Sense-and-Respond framework. We also discuss the monitoring framework that identifies the functional components necessary for real-time monitoring, analysis and optimization of business operations and the supporting IT infrastructure.

12.4.1 Model-Driven Architecture

The linkage of business and IT models in a multi-level model has the potential to greatly reduce the time-to-value of business transformation. This linkage is a significant step towards closing the business-IT gap by maintaining alignment between business design and IT solutions. This linkage also has the potential to provide real-time visibility of business operations which would enable the continual optimization of the business, guided by business-level optimizations and "what-if" analyses.

Model-driven capability design is a core aspect of the Sense and Respond Business Performance Management framework. Instead of automating business processes using workflow management systems and enterprise application integration (EAI) techniques, the model-driven architecture approach is based on building solutions that have ability to respond to changing business conditions (Kumaran 2004). The modeling framework is shown in Figure 12.1.

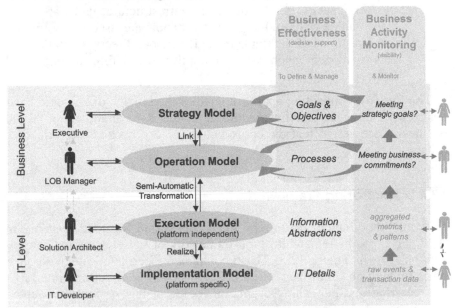

Fig. 12.1. Modeling Framework for a Sense-and-Respond Enterprise

The modeling framework consists of four modeling layers, two layers in the business domain and two layers in the IT domain. An adaptive change at any layer requires validation and verification with higher layers as well as semi-automated propagation of the change to lower layers.

- *Strategy Layer.* The strategy model specifies what the business wants to achieve. It models the business objectives in terms business leaders understand, for example, a description of strategic goals and business objectives in terms of a Balanced Scorecard (Kaplan and Norton 1992).
- *Operation Layer.* The operation model describes what a business is doing to achieve the strategic objectives, and how will it measure progress towards them. It captures the business operations, commitments and key performance indicators (KPIs). The KPIs are directly linked to Balanced Scorecard goals.
- *Execution Layer.* The execution model describes processes and information flows that implement the operation model independent of a particular IT implementation. It is a platform-independent description of documents, flows and their connection to people, applications and data sources.

- *Implementation Layer.* The implementation model defines actual IT processes in a specific realization of the execution model. It is a platform-specific model of the IT infrastructure, hardware, software, middleware and applications. Tools are used today to construct portions of the implementation model directly from the execution model much as a compiler translating a high-level language.

The four-layer modeling approach enables the linkage of strategic business objectives to the IT infrastructure. It increases the alignment of IT and business processes so that the entire enterprise can become performance-driven.

12.4.2 Monitoring Framework

Currently the implementation of many Sense-and-Respond systems in IBM is realized by the architectural framework of loosely coupled components shown in Figure 12.2. These components communicate with each other through an event bus. Each component has well-defined interfaces for receiving and publishing events on the event bus. A loosely-coupled framework gives a Sense-and-Respond designer the freedom to select from a variety of physical components.

The logical components of the framework include:

- *Monitoring Context.* A model that configures and drives Sense-and-Respond activities.
- *Event Emitters.* Placed at appropriate points in the business process and responsible for sending signals and information into the Sense-and-Respond system. This is done by taking a snapshot of key business artifacts and placing a corresponding event on the Event Bus to be consumed by other components in the Sense-and-Respond system.
- *Event Bus.* The central component of the architecture. Other components publish events on the bus and consume events placed on the bus. Raw events published on the bus by Event Emitters are consumed by Business & IT Event Correlation Engines, which calculate KPIs and check for situations, which are either exceptions or noteworthy trends. Situations are published on the bus and consumed by Business Effectiveness Agents. Decisions made by Business Effectiveness Agents are published on the bus and communicated to users in Business Activity Workplaces.
- *Business & IT Event Correlation Engines.* Receive raw events published by Event Emitters. Correlation Engines parse each

event, correlate multiple events, perform complex aggregations and recalculate KPI's from the data contained in events. KPIs are stored in the Business Data Store while events are stored in the Event Store. Correlation Engines evaluate new KPI values against predefined commitments (e.g. KPI thresholds) and publish situations if any commitments have been violated. Correlation Engines also try to detect important trends that could lead to violated commitments in the future.

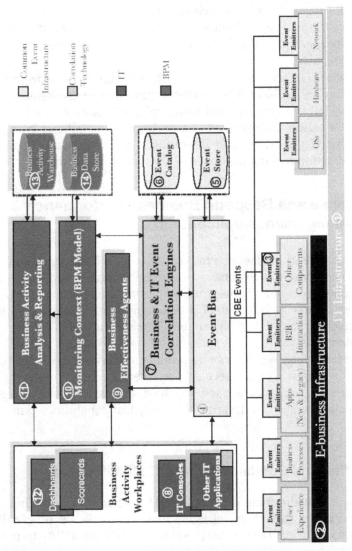

Fig. 12.2. Architectural framework for Sense-and-Respond business management

- *Business Effectiveness Agents*. Receive situations published on the Event Bus and propose one or more actions. Actions can fall into a number of categories, including notifications to key business managers, changes to operational parameters or business rules, re-allocation of resources, invocation of exception processes, improvement of ineffective processes and improvement of ineffective strategies.
- *Business Activity Analysis and Reporting*. Utilizes data in the Business Activity Warehouse to support trend analysis and root cause analysis. Standard OLAP (On Line Analytic Processing) tools are provided for analysis, as well as advanced visualization graphics.
- *Business Activity Workplaces*. Receive information from the Event Bus and present it to business users in various formats. Provide visibility to current and historical KPI values as well as trend information. Support root cause analysis, in concert with Business Activity Analysis and Reporting. When situations arise, support decision making in concert with Business Effectiveness Agents.

12.5 Sense and Respond Business Performance Management Applications

Over the course of the last several years, we have built Sense and Respond systems and consulted customers in various industries and to help manage high-technology value chains, transportation management logistics, retail and service parts logistics, steel production, and banking operations. All these projects share the same reference architecture and solution patterns and components.

In the following, we describe two pilots to illustrate how the Sense and Respond Business Performance Management framework was applied to each scenario. Analytics are the key to a successful Sense-and-Respond implementation, and we describe the analytical capabilities built for both systems.

12.5.1 Demand Conditioning for PC Manufacturing

The demand conditioning process at IBM's Personal Computing Division (PCD) began as part of an initiative to improve the on-time delivery of PCD products to customer orders, and to improve the ability to predict and respond to supply and demand imbalances. The goal of the pilot was to enhance supply chain visibility and proactively develop a better understanding of transactional data representing customers' needs. To support this process, we developed a web-based Sense and Respond system that identifies order events and available supply headlights across the order-to-delivery supply chain. The system monitors supply and demand imbalances for commodities, and indicates out-of-threshold situations on an enterprise dashboard. A key innovation in this pilot is a new algorithm that identifies potential gaps by using historical information and future indicators to forecast ordering trends. The new algorithm has been coupled with improved data integration and a web-based management dashboard that provides a current view of key supply and demand metrics for each IBM PC component.

The principals of demand conditioning are threefold, involving the supply of commodities, the product offering and sales plan as shown in Figure 12.3.

Each of these three areas provides unique capabilities to make the conditioning process work.

- *Procurement supply conditioning.* Focuses on working with suppliers to improve flexibility in supply to react to customer demand that is never totally predictable.
- *Offering conditioning.* Focuses on identifying alternative products or substituting PC components that can be provided to customers in reaction to supply imbalances. It is supported by a proactive product definition phase that provides more flexibility to define product configurations.
- *Demand conditioning.* Focuses on providing a dynamic sales plan in the sense that it can be changed in reaction to supply imbalances. It considers pricing actions and promotions to provide incentives to customers to choose alternatives.

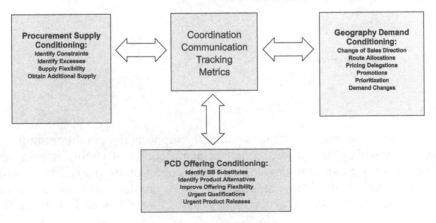

Fig. 12.3. Principles of Demand and Supply Conditioning

The PCD Demand Conditioning Process illustrated in Figure 12.4. provides a management system with which to apply these three principals. The process involves people in different organizations at locations worldwide. The execution of the process revolves around a weekly core team meeting led by PCD's Worldwide Fulfillment Organization (WWFO). The team consists of representatives from the PCD Brand, Operations, Procurement, Finance and Product Development. This team identifies supply imbalances, creates a conditioning plan in partnership with the geography sales organizations, and manages the execution of the conditioning plan. As the solution is executed, the actions taken in the three principals are being tracked to ensure that the solution is being executed properly. Finally, metrics involving customer orders must also be tracked to make sure that the solution is being effective. This process was begun in August 2003, and since that time several supply imbalances have been successfully conditioned.

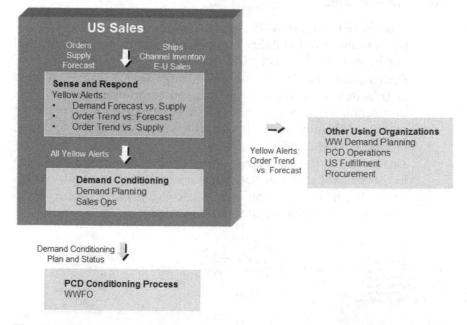

Fig. 12.4. PCD Demand Conditioning Process

The Sense and Respond system directly supports the conditioning process by providing an earlier, proactive identification of supply imbalances that makes it possible to develop effective conditioning plans. The system receives daily order loads, shipments, supply commits and demand forecasts from PCD enterprise planning systems, correlates and analyzes the information, alerts the appropriate business users and recommends corrective actions.

Action tracking capabilities are provided which record a snapshot of the data at the time the alert was generated and compares it to the current over a pre-determined time horizon. This provides benefits in two respects:

- Providing a capability to track the performance of the actions which were invoked in response to a business exception.
- Building a rich history of actions in the data warehouse over which intelligent mining operations can be performed to learn and recommend actions in the future.

A key innovation in this pilot is a predictive analysis of orders that aims at developing short-term visibility (typically 4-6 weeks) into customer ordering behavior as an early indicator of supply imbalances. The order trend analysis is utilized to compare trends to the demand forecast as a lead indicator of future supply imbalances. Part of the weekly review is to select technologies where this indicator shows a potential issue and review the forecast with the US planning team.

Unlike long-range forecasting techniques, the order trend analysis identifies repetitive historical patterns of orders, and obtains accurate short-term predictions of order rates through increased consideration of data available in order execution systems. Coupled with improved data integration and the web-based management dashboard, the order trend analysis enables a current view of key supply and demand metrics for each IBM PC component.

The order trend analysis is based on a model that utilizes historical and future demand-related indicators such as actual demand and customer order inflow. The model estimates the effects of seasonality, order skew within a quarter, product life cycles, and repetitive order trends from historical data. It also provides point estimates, percentiles and confidence intervals for risk management. The order trend analysis combines traditional statistical forecasting techniques with demand-related indicators visible in the current time period that can serve as headlights for future demand to improve baseline forecasts. The order trends are operational forecasts that provide a more accurate picture of demand for the next 4-6 weeks which is the most critical time for deployment.

The three indicators that were integrated into the analysis are:

- *Total order load.* The current amount of unfilled customer orders with a customer requested shipment date some time in the future.
- *Order coverage.* The current amount of supply-committed customer orders with a confirmed future shipment date.

–   *Channel inventory.* The current amount of inventory stocked at a business partner's warehouses to fill future customer demand.

The order trend analytics are executed on a daily basis in the Sense-and-Respond system to produce new order trends. During the initial deployment of the Sense-and-Respond technology, the algorithm has already proven to be much better at predicting actual future orders. Part of the weekly conditioning process is to select technologies where this indicator shows a potential issue, and review the forecast with the US geo planning team.

Figure 12.5 illustrates the linkages between the Sense-and-Respond system and PCD data sources and supply chain applications.

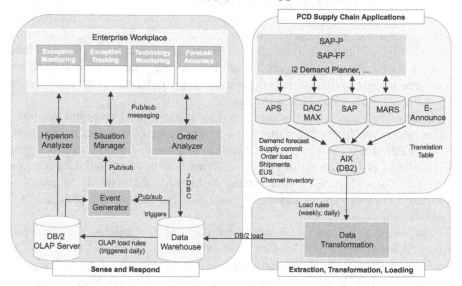

Fig. 12.5. PCD Sense-and-Respond Architecture

The Data Extraction, Transformation, and Loading component (ETL) accesses transactional tables in PCD supply chain planning and execution systems. These contain demand forecasts, supply commits, order loads, order shipments, end-user sales and business partner inventory. The load rules also explode the source data (provided at the fully configured system level) to the PC component level via a product-to-technology translation table which was a key requirement for the pilot.

The Observation Manager provides the functions for correlating and analyzing transactional data to detect business exceptions by comparing supply and demand indicators over a rolling time horizon. The business rules utilize cumulative differences between supply and demand which is the basis for detecting supply shortages or supply overages.

A relational Data Warehouse captures the order loads, shipments and planning data at a system level as well as component level. It is augmented by business intelligence that provides root-cause analysis by connecting an OLAP client to the dashboard.

The Enterprise Workplace provides an end-to-end view of the imbalances between supply and demand to enable successful conditioning. It allows for customization and administration by the different role players. It also recommends actions based on alerts generated and provides capabilities to track the actions thereby enabling business effectiveness. Figure 12.6 shows the main screen of the Enterprise Workplace.

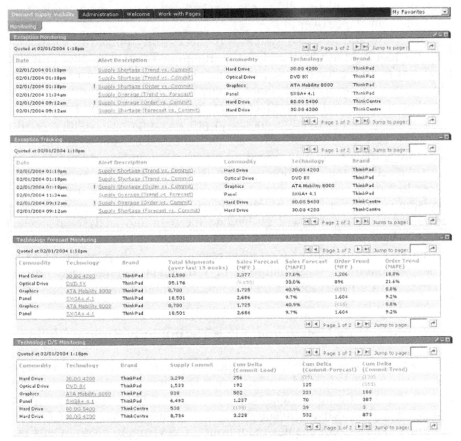

Fig. 12.6. Top-Level View of the Sense and Respond Enterprise Workplace

The top-level screen consists of four sections. The Exception Monitoring Portlet displays all current alerts for potential overages and underage situations, for example when the order trend is significantly departing from the demand forecast. The Exception Tracking Portlet allows business users to record conditioning actions and monitor the impact and effectiveness of executing recommended actions. The Forecast Monitoring Portlet shows the historical accuracy of the planned demand forecast and the order trend as measured by the mean forecast error (MFE) and the mean absolute percentage error (MAPE). Finally, the Technology Monitoring Portlet shows the latest demand and supply status of key PC components.

Figure 12.7 shows a sample detail view of a supply shortage alert, in this example for a hard drive component. The order trend is displayed for the future thirteen weeks in the form of weekly point forecasts and their confidence intervals. The order trend is compared to the latest supply projection, and a cumulative difference is calculated to track the amount of the projected imbalance over time.

The Sense and Respond system was successfully piloted in early 2004 and the tool is in use by the PCD Conditioning Team. The pilot provided business performance benefits, enabling sales teams to adjust selling tactics and supply teams to rebalance supply more quickly and effectively. A number of functional enhancements are currently under way. First, we are enhancing the order trending model to provide improved volume predictions for new product introductions and end-of-life situations based on technology transitions maps. Second, we are developing capabilities to record the actions the conditioning team is taking to resolve supply imbalances to form a knowledge base for data mining that will be used to assist decision making for future supply imbalances. And third, we are building advanced analytics that will go beyond the data and analysis associated with demand planning, extending into lower tier suppliers and optimizing inventory hubs and buffers for a more responsive supply pipeline. The analytics will facilitate the monitoring of fulfillment activities and provide metrics and alerts that focus attention on serviceability issues.

12.5.2 Inventory Management in a Technology Supply Chain

The next pilot was developed to support an automated inventory management process at IBM's Microelectronics Division (IMD). This pilot utilizes Sense-and-Respond capabilities to improve internal business processes via KPIs such as inventory turns, on-time delivery, and forecast accuracy. The system enables monitoring of key supply chain events that help manage the supply chain's performance and to achieve customer service requirements with the minimum possible inventory. We piloted the tool successfully with the IMD inventory planning team at the end of 2003.

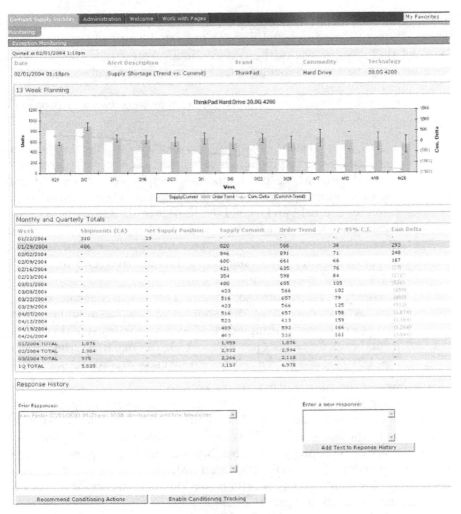

Fig. 12.7. Detail View of a Supply Shortage Alert

A key ingredient of the system is the analytical model that optimizes inventory positioning in the IMD semiconductor supply chain. The analytics complement existing planning applications by leveraging transactional data from enterprise business applications. Through this technology, business managers at IMD are able to make adjustments to optimize inventory, based on monitored performance and to reduce the response time by using decision analysis support.

IMD business managers had been looking for ways to improve operational performance and reduce expenses associated with inaccurate Original Equipment Manufacturer (OEM) forecasts, inefficient order flow, expedited shipments and obsolete inventory. Like many other organizations,

IMD is faced with the challenge of responding and adjusting to supply chain events in a synchronized, timely, and intelligent fashion. They recognized that the key was to have a continuous process of performance measurements that would identify problem areas in the end-to-end supply chain on a timely basis.

Within the IMD end-to-end supply chain there were two key processes requiring response:

- *Supply management* - focusing on changes or modifications in work-in-process or production parameters such as yields and cycle times. Sense-and-Respond monitors supply versus demand and capacity utilization to provide key reports when demand is in jeopardy and help understand the impact of the tardiness. This information enables analysts to gauge anticipated supply against demand by demand class, immediate identification of demands in jeopardy, identify assets supporting this demand, and full profile of anticipated capacity utilization.

- *Inventory management* - focusing on changes in a business policy such as inventory days of supply which is also impacted by changes in manufacturing practice such a shorter cycle times. The inventory management process controls the manufacturing of wafers, devices, and modules based on inventory reorder points.

To improve the above management processes, we first developed an analytical supply chain model that optimizes target inventory levels at different stages of manufacturing. The model helps to identify potential shortages of finished goods and avoid obsolescence and delinquent customer deliveries. This analytical capability was the key to proactive business management. The model improved IMD's inventory management process by diagnosing supply shortfalls, backlog accumulation, and inadequate inventory levels at strategic stocking points.

The analytical supply chain model was then combined with Sense-and-Respond performance management applications to enables pro-active exception detection by monitoring customer demand, inventory and shipments relative to predefined objectives. When performance metrics go outside of acceptable limits, the applications automatically alert inventory planners so they can investigate the issue.

A relational data warehouse that contains up-to-date profiles of business metrics for event engine processing serves as the primary data repository for event trails from enterprise applications and advanced planning systems. The data warehouse also contains operational manufacturing parameters such as bills of materials, lead times, process yields, demand forecasts, and supply commits that are used as inputs to the inventory optimization module. The Sense-and-Respond system retrieves transactions and planning data from enterprise planning and execution systems. The transactional data is organized and stored in a data warehouse from which metrics and KPIs are calculated.

The inventory optimization module provides business intelligence and analytics to improve the performance of the enterprise. It adopts existing business processes and cost structures, and recommends optimized operational inventory policies that drive business performance to higher levels of operational and financial efficiency. The recommendations allow business process owners to see the expected impact of planning decisions, assess the profit risk and rewards of proposed actions, and evaluate alternative options. The optimization model consists of a three-echelon structure with an additional assembly node, including wafer fabrication, wafer test, substrates and bond-assembly and test. The objective is to minimize inventory subject to a service requirement measured as on-time delivery to customers within an allowed lead time window. Figure 12.8 is a graphical illustration of the model and the solution approach.

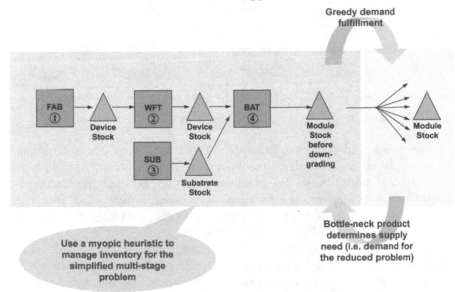

Fig. 12.8. High-Level Illustration of the Analytical Inventory Model

The model utilizes demand forecasts, manufacturing cycle times, yields, costs, lot sizes, inventory policies, contractual buffers, customer service targets, product prices, and the rates of change in prices and costs. Based on all these input parameters, it calculates and reports operational and financial performance for business managers and inventory planners. The performance reports comprise numerous financial and operational performance metrics as illustrated in Figure 12.9. These metrics are projected for several weeks or months into the future. The analytical model also determines optimal operational days of supply policies at strategic stocking locations in wafer fabrication and module assembly and testing plants.

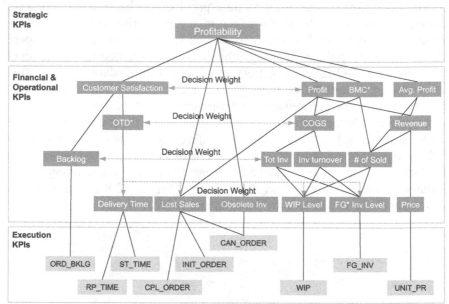

*FG: Finished Goods, COGS: Costs of Goods Sold, OTD: On-Time Delivery, BMC: Base Manufacturing Cost, WIP: Work in Process

Fig. 12.9. Financial and Operational Metrics Reported by the Inventory Model

The Sense and Respond Enterprise Workplace provides a dashboard view of the overall health of the business. The dashboard is role-based with distinct portal views for inventory analysts, product line managers, supply chain executives, and financial executives. Figure 12.10 illustrates the visibility screen detailing the inventory status, customer delivery performance and order fulfillment related metrics.

The Enterprise Workplace also supports what-if analyses to evaluate the impact of various manufacturing and demand characteristics on inventory turns and customer service levels. Users can view demand forecasts, manufacturing cycle times, yields, costs, lot sizes, inventory policies, contractual buffers, customer service targets, product prices, and the rates of change in prices and costs. Figure 12.11 shows the what-if analysis view of the scenario management portlet.

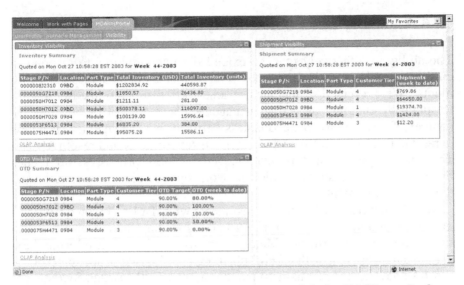

Fig. 12.10. Visibility Portlet View for Inventory and Order Fulfillment Performance

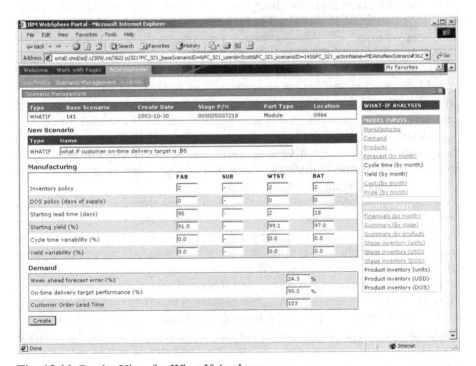

Fig. 12.11. Portlet View for What-If Analyses

12.6 Summary and Future Study

The ability to effectively manage extended value networks to respond to customer needs is critical in today's rapidly changing business environments. We presented a technical framework that supports Sense-and-Respond by enabling proactive management and control of business resources. Sense and Respond Business Performance Management is a new paradigm that integrates real-time decision support, risk and resource management, supply chain optimization, and business processes. It blends business and IT to support value network optimization in uncertain and dynamic environments.

We also described two pilot engagements with IBM's Personal Computing Division and IBM's Microelectronics Division where Sense and Respond Business Performance Management was applied. Our pilots and experience indicate that with careful planning and focused scope, enterprises can take advantages of this new paradigm, even with partial, incremental implementation on some of the capabilities discussed.

The technologies discussed in this chapter are either available today, or are emerging. However, the integration of these technologies still poses significant challenges that need to be addressed. For example, the innovative applications of Web-Services, Component-Based Modeling, and Rapid Integration of these technologies in the Sense and Respond framework need further studies. Another critical issue is the data availability and reliability. Businesses are normally reluctant to share sensitive data. And the reliability of data, about people, organizations, customers and partners can affect value network decisions. Support for trust-enabled value networks and technologies to help assess reliability and third party services to filter and aggregate sensitive data can fundamentally change how value networks are formed, interact, share knowledge, use information, and make decisions (Wang and Todd 2003).

In addition to process and technology, the transformation into an integrated Sense and Respond enterprise requires a mastery of organizational complexity and culture. Transforming business to enable proactive Sense and Respond capabilities require fundamental changes in how people manage information, collaborate, and use the technologies. The cultural and transformational aspects of Sense and Respond are critical success factors in implementation, yet they are often neglected by efforts seeking real-time responses to events. Hence, additional investigation and research about business ecosystems and the social aspects of participants, and their effects on value networks are needed. We believe the early adopters, who can effectively manage the technologies and organizational changes needed for enabling the proactive Sense and Respond capabilities in an incremental and transformational approach will enjoy significant competitive advantages.

References

Bittner, E. (2000) E-Business Requires Supply Chain Event Management, AMR Research

Bradley, S., Richard, P., Nolan, L. (eds.) (1998) *Sense and Respond: Capturing Value in the Network Era*, HBS Press Book

Correia, J., Schroder, N. (2002) BAM: A Composite Market View, Gartner Inc.

Haeckel, S. (1999) *Adaptive Enterprise: Creating and Leading Sense-and-Respond Organizations*, Harvard Business School Press, Boston, MA

IBM (2003) Adaptive e-Business: Enabling the On-Demand Enterprise. IBM Global Services White Paper

Kaplan, R., Norton, D. (1992) The Balanced Scorecard – Measures That Drive Performance, *Harvard Business Review*

Kumaran, S. (2004) Model Driven Enterprise, Global EAI Summit, Banff, Canada

Lawrie, G. (2003) Preparing for Adaptive Supply Networks, Forrester

Lee, H., Amaral, J. (2002) Continuous and Sustainable Improvement Through Supply Chain Performance Management, *Stanford Global Supply Chain Forum*

Lehmann, C. (2003) The Rapid Sense-and-Respond Enterprise: Part 1 & Part 2, Meta Group Reports

Lin, G., Ettl, M., Buckley, S., Bagchi, S., Yao, D.D., Naccarato, B.L., Allan, R., Kim, K., Koenig, L. (2000) Extended-Enterprise Supply-Chain Management at IBM Personal Systems Group and Other Divisions, *Interfaces* 30 (1): 7–21

Lin, G., Buckley, S., Cao, H., Caswell, N., Ettl, M., Kapoor, S., Koenig, L., Katircioglu, K., Nigam, A., Ramachandran, B., Wang, K.-Y. (2002) The Sense and Respond Enterprise, *OR/MS Today* 29 (2): 34–39

Lin, G., Jeng, J.J., Wang, K. (2004) Enabling Value Net Collaboration, in: *Evaluation of Supply Chain Management*, Chang, Y.S., Makatsoris, H.C.,Richards, H.S. (eds.), Kluwer Academic Publishers, 417–430

Suleski, J., Quirk., C. (2001) Supply Chain Event Management: The Antidote for Next Year's Supply Chain Pain, AMR Research

Wang, K, and Todd, A. (2003) Trust Enabled Business at a Glance. IBM Global Services Enterprise of the Future White Paper

13 Authors' Profiles

Mark Adelhelm *IBM Business Consulting Services, 8623 Great Cove Drive, Orlando, FL 32819 (adelhelm@us.ibm.com)*. Mr. Adelhelm is an Associate Partner and leader of IBM's Supply Chain Strategy competency in the Distribution Sector, where he serves Retail, Consumer Products, Life Science and Travel-related clients. Mark has 15+ years experience in supply chain and operations management with particular focus in strategy development, enterprise transformation, advanced supply chain planning and customer order management and fulfillment solutions. He received a B.S. in Industrial and Systems Engineering from Georgia Tech and an MBA from the University of Chicago with concentrations in Finance and Corporate Strategy.

Chae An *IBM Software Group, 294 Route 100, Somers, NY 10589 (chaean@us.ibm.com)*. As the Director of Industrial Sector Solutions for IBM Software Group, Dr. An is responsible for creating software solutions that address the core business problems for IBM's customers in automotive, electronics & electrical, aerospace, and chemical & petroleum industries. Prior to his current position, he was the Director of e-Commerce Research at IBM Thomas J. Watson Research Center. At IBM Research, he was responsible for leading research activities in the areas of supply chain management, e-Commerce technologies, e-marketplaces, business process integration, and advanced decision support and dynamic pricing algorithms. His experiences in IBM also include serving as the solution executive for the Supply Chain Optimization Solutions unit, during which he ed the development, services, and sales of the manufacturing supply chain solutions for IBM. He received his B.S., M.S., and Ph.D. in Electrical Engineering from MIT. He is a member of IEEE and INFORMS.

Sugato Bagchi *IBM Research Division, Thomas J. Watson Research Center, P.O. Box 218, Yorktown Heights, NY 10598 (bagchi@us.ibm.com)*. Dr. Bagchi is a Research Staff Member at the IBM Thomas J. Watson Research Center in Yorktown Heights, NY. His research interest is in financial valuation of information technology applications and infrastructure. He has developed business value models as well as their underlying tools and techniques to help potential adopters of emerging technologies quantify the predicted benefits in terms of business and financial impact. In this role, he has assisted several IBM Global Services engagements with the task of developing business cases for their clients. Previously, Dr. Bagchi worked on knowledge representation frameworks and methodologies for business strategy formulation and analysis. This work is being done in collaboration with IBM e-business strategy consultants. He has contributed subject matter expertise to numerous strategy consulting engagements. In the recent past, Dr. Bagchi has worked on the design and implementation of a business process modeling and simulation tool, as well as a specialization of this tool for multi-enterprise supply chains. Dr. Bagchi has a Ph.D. degree in Electrical and Computer Engineering from Vanderbilt University, where his thesis was on task planning under uncertainty.

Kumar Bhaskaran *IBM Research Division, Thomas J. Watson Research Center, P.O. Box 218, Yorktown Heights, NY 10598 (bha@us.ibm.com).* Dr. Bhaskaran has been actively involved for the past fifteen years in the architecture, design, and development of supply chain and enterprise business solutions. He is currently leading research in the area of service-oriented architectures and technologies for business integration and business performance management solutions. Kumar has a Ph.D. in Industrial Engineering & Operations Research from the Rensselaer Polytechnic Institute, Troy, NY.

Steve Buckley *IBM Research Division, Thomas J. Watson Research Center, P.O. Box 218, Yorktown Heights, NY 10598 (sbuckley@us.ibm.com).* Steve Buckley has been a Research Staff Member at the IBM Thomas J. Watson Research Center in Yorktown Heights, NY since 1987, and a manager at that facility since 1995. He currently manages the Analytic Models & Architecture department in the Mathematical Sciences organization. His most recent interest is in Sense-and-Respond systems. His team has implemented Sense-and-Respond for several IBM lines of business. He also has a continuing interest in supply chain simulation and optimization. Dr. Buckley received the Ph.D. degree in Computer Science from MIT in 1987. He also received the M.S. degree in Computer Science from Penn State in 1978, and the B.S. degree in Applied Mathematics and Computer Science from Florida State in 1977.

Feng Cheng *IBM Research Division, Thomas J. Watson Research Center, P.O. Box 218, Yorktown Heights, NY 10598 (fcheng@us.ibm.com).* Dr. Cheng is a Research Staff Member at the IBM Thomas J. Watson Research Center. He joined IBM Research Division in 1996 after receiving a Ph.D. degree in Operations Management from the University of Toronto. His current research areas include business process value modeling, supply chain simulation and optimization, and operational risk management.

Daniel Connors *IBM Research Division, Thomas J. Watson Research Center, P.O. Box 218, Yorktown Heights, NY 10598 (dconnors@ us.ibm.com).* Daniel Connors received the B.S.E. degree in Electrical Engineering from the University of Michigan in 1982, the M.S. and Ph. D. degrees in Electrical Engineering from the University of Illinois in 1984 and 1988, respectively. Since 1988, he has been a research staff member at the IBM Thomas J. Watson Research Center. His research interests include modeling, simulation and design of business processes and design of decision support tools for manufacturing and supply chain logistics. He is in the Mathematical Sciences Department in IBM Research where he works on optimization for professional services automation.

Brenda Dietrich *IBM Research Division, Thomas J. Watson Research Center, P.O. Box 218, Yorktown Heights, NY 10598 (dietric@us.ibm.com).* Dr. Dietrich is the Director of the Mathematical Sciences department of the IBM Thomas J. Watson Research Center, Yorktown Heights, NY. Previously, she was the senior manager of the Optimization Center at IBM Research, where she managed both the optimization research and the application of optimization to supply chain and transportation. She founded the Logistics Applications group and was the IBM Research liaison for travel and transportation projects. In 1995 she served on the Research Technical Planning Staff, coordinating the development of a technology outlook and portions of the division strategy. From 1990 to 1994 she lead the development of a set of resource allocation and planning tools for IBM manufacturing lines, and managed the Manufacturing Logistics group, and the Manufacturing Planning and Scheduling group at IBM Research. Her research includes work in manufacturing modeling and scheduling, inventory management, transportation logistics, mathematical programming, and combinatorial optimization. She is a member of the Advisory Board of the Department of Industrial Engineering and Management Sciences (IE/MS) at Northwestern University, of IMA at University of Minnesota, and DIMACS at Rutgers, a member of the IBM Academy of Technology, and has served on the Board of INFORMS. She has co-authored twelve patents and numerous publications, and recently co-edited a book on the mathematics of internet auctions. Dr. Dietrich joined IBM Research in 1984 and holds a Ph.D. in Operations Research and Industrial Engineering from Cornell University

Michel W.F.M. Draper *IBM Global Services, David Ricardostraat 2-4, 1066JS Amsterdam, The Netherlands (draper@nl.ibm.com).* Michel Draper is a consulting IT Specialist in the area of Business Analysis at IBM Application Maintenance Services in Amsterdam, The Netherlands. He is working within internal as well as external projects usually within a team in the role as Business Information Analyst or Business Architect.

Prior to this appointment he was a senior IT Specialist at EMEA (Europe, Middle East and Africa) Service Logistics (1995-2004). EMEA Service Logistics is part of IBM's Global Service Logistics unit. Global Service Logistics develops, implements and manages service parts logistics solutions and systems for IBM and other parties usually at a global scale. Before joining IBM he worked at KLM Royal Dutch Airlines as a consultant Decision Support systems (1988-1995). He joined various projects as a consultant in the airline industry. Previously he worked at the Free University of Amsterdam teaching Management Information Systems and Operations Research (1986-1988). In 1986 he completed his master of Econometrics with specialization on Operations Research at the Free University in Amsterdam.

Thomas Ervolina *IBM Research Division, Thomas J. Watson Research Center, P.O. Box 218, Yorktown Heights, NY 10598 (ervolina@ us.ibm.com).* Dr. Ervolina joined IBM in 1989 and has been a been a Research Staff Member at the IBM Thomas J. Watson Research Center in Yorktown Heights, NY since 1993. His major work and interest is in the creation of algorithms and software for Resource Allocation in a Supply Chain. He has developed optimization software for Available to Promise and Supply/Demand Planning within an extended supply chain. He works closely with IBM's internal supply chain operations to integrate, deploy and support the software in an environment that must frequently adapt to new business environments. He received his Ph.D. degree in Operations Research from Columbia University in 1989. His thesis work was in the areas of Combinatorial Optimization and Network Flows. He also holds an M.S. degree in Operations Research from Columbia, and a B.S. degree in Mathematics from SUNY Stony Brook.

Markus Ettl *IBM T.J. Watson Research Division, P.O. Box 218, Yorktown Heights, NY 10598 (msettl@us.ibm.com).* Dr. Ettl is a Research Staff Member at IBM's T.J. Watson Research Center. He joined IBM Research in 1995 after receiving his doctoral degree in Computer Science in 1995 from Friedrich-Alexander University in Erlangen, Germany. Dr. Ettl's research interests span several applied areas including business process modeling, simulation, and design of decision support systems for manufacturing logistics. His current primary research interest is in Sense-and-Respond systems. He is co-author of IBM's work on extended-enterprise supply chain management, which was awarded the INFORMS Franz Edelman Award in 1999.

J.P. Fasano *IBM Research Division, Thomas J. Watson Research Center, P.O. Box 218, Yorktown Heights, NY 10598 (jpfasano@us.ibm.com).* J.P. Fasano is the manager of Optimization and Mathematical Software at the

IBM Thomas J. Watson Research Center, in Yorktown Heights, NY. He has degrees in System Engineering, Computer Science, and Operations Research, all from Rensselaer Polytechnic Institute. His research activities and interests include applications of optimization to supply chain management.

Moritz Fleischmann *Rotterdam School of Management, Erasmus University Rotterdam, P.O. Box 1738, 3000 DR Rotterdam, The Netherlands (MFleischmann@fbk.eur.nl).* Dr. Fleischmann is an Assistant Professor of Quantitative Methods at the Department of Decision and Information Sciences of the Rotterdam School of Management. He received his M.S. in Business Mathematics from the University of Ulm (Germany) in 1996 and his PhD in General Management from Erasmus University Rotterdam in 2000. In 2002, he spent a year as Visiting Assistant Professor at the Tuck School of Business at Dartmouth College (New Hampshire). Dr. Fleischmann's research addresses topics in the field of supply chain management. Besides reverse logistics, current focal points are on coordination of pricing and operations, and on multi-channel distribution, in particular e-fulfillment.

Hansjörg Fromm *IBM Business Consulting Services, Pascalstr. 100, 70569 Stuttgart, Germany (fromm@de.ibm.com).* Dr. Fromm studied Mathematics and Computer Science and received his Ph.D. from the University of Erlangen-Nürnberg, Germany, in 1982. After a research assignment at the IBM Thomas J. Watson Research Center, Yorktown Heights, NY, he joined IBM Germany in 1983, where he assumed different management positions in software development, quality assurance, and manufacturing research. After joining IBM Global Services in 1995, Dr. Fromm was a pioneer in building up a Supply Chain Management competency in Europe. He initiated and led successful SCM projects mainly in the Automotive Industry. Today, he is a Partner of IBM Business Consulting Services with focus on Automotive Supply Chain and Spare Parts Management. Dr. Fromm is a member of the IBM Academy of Technology and an IBM Distinguished Engineer. He is a honorary professor at the University of Erlangen-Nürnberg, where he teaches on Supply Chain Management and e-Marketplaces.

Guillermo Gallego *Columbia University, Industrial Engineering and Operations Research, 324 Mudd Bldg, 500 West 120th Street, New York, NY 10027 (gmg2@columbia.edu).* Dr. Gallego is a Professor and the Department Chair of the Industrial Engineering and Operations Research Department at Columbia University. He received a B. A. degree in mathematics from the University of California at San Diego, and M. S. and Ph. D. degrees from Cornell University in 1986 and 1988, respectively. He subsequently joined Columbia University where he has worked on supply chain management and revenue management. Professor Gallego worked at the IBM Thomas J. Watson Research Center as a visiting scientist from 1999-2003.

Rainer Gapp *IBM Global Logistics, Wilhelm-Fay-Str. 30-34, 65936 Frankfurt, Germany (rainer_gapp@de.ibm.com).* Mr. Gapp is program manager for Logistics Operations within IBM's Global Asset Recovery Services organization for EMEA. This responsibility covers overall EMEA reverse logistics including technical operations such as remanufacturing, demanufacturing and scrap performed in the Montpellier (France) and Mainz (Germany) centers. After receiving his degree in computer science from the University of Hildesheim, he started his career within IBM in 1997. Prior to his current position, he was a customer engineer for hardware and software and held various team leader positions in the EMEA remanufacturing center of IBM in Germany. He has been working on several EMEA return process optimization projects since 1999.

Ben Gräve *IBM Global Services, Transistorstraat 7, 1322CJ Almere, The Netherlands (graeve@nl.ibm.com).* Ing. Gräve is the global executive of the IBM After-Sales Service Parts Supply Chain. After his degree in electronics engineering he worked for Philips Telecommunication as a development engineer. Since 1974, when he joined IBM, he has held various positions in Manufacturing Engineering and Materials Management in The Netherlands, United States and United Kingdom. As part of IBM Global Services he has led major re-engineering efforts in Europe resulting in a fully integrated pan-European Forward and Reverse Service Parts Logistics business model. Since he became the Director of Global Service Logistics, he is leading the transformation of the various geographies into an on demand global operating model. As a leading strategist in his field, he is also board member of various logistics institutes in Europe where different industries and universities meet to address strategic logistic issues.

William Grey *IBM Retirement Funds, 1133 Westchester Avenue, White Plains, NY 10604 (wgrey@us.ibm.com).* William Grey is currently an Investment Advisor in the Risk Management department of IBM Retirement Funds. Mr. Grey spent most of the last decade at the IBM Thomas J. Watson Research Center, where he conducted research on information technology valuation and risk management, developed innovative techniques for managing high technology supply chains, and designed decision support tools to improve business performance. His most recent Research position was Program Director for Value and Risk Modeling. In that role, he led a team that developed analytic tools and methodologies to support new consulting offerings to quantify the business impact and risk of information technology investments.

Robert Guttman *IBM Research Division, Thomas J. Watson Research Center, P.O. Box 218, Yorktown Heights, NY 10598 (rguttmann@ us.ibm.com).* Mr. Guttman leads a worldwide initiative within IBM T.J. Watson Research Center which prescribes a model-driven approach to translate a business' strategic intent into a deployed implementation. Upon receiving his MS in 1998 from MIT, Rob founded Frictionless Commerce, a leading enterprise sourcing software vendor, based on software agent technologies he invented at MIT's Media Laboratory. These technologies are automating sourcing business processes, providing visibility into sourcing activities and information through role-based dashboards, and helping procurement professionals make optimal award allocation decisions resulting in direct, ongoing bottom-line savings. These innovations earned Rob a 2002 MIT Technology Review Top 100 Young Innovators Award (TR100). Rob received a BSE in Computer Engineering from the University of Michigan in 1992.

Aliza Heching *IBM Research Division, Thomas J. Watson Research Center, P.O. Box 218, Yorktown Heights, NY 10598 (ahechi@us.ibm.com).* Dr. Heching is a member of the Business Analytics and Optimization group at the IBM T.J. Watson Research Center. She received her PhD in Management Science/Operations Research from the Columbia University Graduate School of Business. Her doctoral work focused on combined pricing and inventory control strategies for retailers. Dr. Heching joined IBM Research in 1998. She is currently working on a number of business optimization problems including optimal design of supply contracts, ROI and business value modeling, service-after-sales planning, and planning for corporate IT spending.

Jayant Kalagnanam *IBM Research Division, Thomas J. Watson Research Center, P.O. Box 218, Yorktown Heights, NY 10598 (jayant@us.ibm.com).* Dr. Kalagnanam is a research staff member at IBM Watson since 1996. He received his Ph.D. from Carnegie Mellon University in 1991. His research interests lie in the general area of decision support, optimization, economics and their applications to Business Analytics and Optimization. Since joining IBM research he has been involved in real-world projects where he deployed optimization and decision support solutions. One set of projects are focused on production planning and operations scheduling (www.research.ibm.com/pdos) and another set of projects involve procurement and sourcing (www.research.ibm.com/auctions).

Kaan Katircioglu *IBM Research Division, Thomas J. Watson Research Center, P.O. Box 218, Yorktown Heights, NY 10598 (kaan@us.ibm.com).*

Dr. Katircioglu is one of the leading researchers and an ODIS consultant in supply chain management at IBM Corporation. He has more than ten years of experience in the field of Operations Research, Management Science and Logistics. His expertise covers the areas of inventory optimization, distribution and manufacturing operations management and planning, e-business and supply chain management. He has a bachelor's degree in Industrial Engineering and a Master's degree in Statistics. He completed his Ph.D. in Management Science / Transportation / Logistics at the University of British Columbia in 1996. While pursuing his doctoral studies, he worked at UBC as a research assistant from 1991 to 1996. He also taught courses in business statistics and quantitative methods in business. He joined IBM as a researcher at T.J. Watson Research Center in 1996. Since then, he has worked on several projects for various divisions of IBM and its customers, published papers, and made several conference presentations. He is a member of INFORMS, IEEE.

Colin Kessinger *Vivecon Corporation, 650 Castro Street, Suite 300, Mountain View, CA 94041 (colin.kessinger@vivecon.com).* Colin Kessinger is one of the founders of the Vivecon Corporation, a pioneer in delivering Supply Chain Risk Management solutions. The technology integrates the tools and principles of financial engineering and supply chain management to quantify performance risk across three key dimensions: cost, availability, and liability. Prior to founding Vivecon, Dr. Kessinger received his PhD from Stanford University (Industrial Engineering) and was a professor of Operations Management at the University of Michigan. During this period, his research focused on supply risk and flexibility management and the impact of supply contracts on optimally allocating risk throughout the supply chain.

Santhosh Kumaran *IBM Research Division, Thomas J. Watson Research Center, P.O. Box 218, Yorktown Heights, NY 10598 (sbk@us.ibm.com).* Dr. Kumaran manages model driven enterprise research team at IBM's T. J. Watson Research Center in Yorktown. He has been with IBM since 1996 and was awarded an IBM Outstanding Technical Achievement award in 2001 and an Execute Now award in 2002. His research interests include business process modeling and model driven business integration and management. He holds a PhD in Computer Science from Oregon State University.

Ying Tat Leung *IBM Research Division, Almaden Research Center, 650 Harry Road, San Jose, CA 95120 (ytl@us.ibm.com).* Dr. Leung recently joined the Services Research Department at the IBM Almaden Research Center as a Research Staff Member, after almost 10 years at the IBM Thomas J. Watson Research Center in New York. He is currently leading a research program in business value modeling. He is also the Research Relationship Manager for the IBM Global Consumer Products Industry, acting as the Research representative for customer activities in this industry. In addition to business value modeling, Ying Tat's technical work spans an entire supply chain: product pricing, demand forecasting, inventory replenishment, production planning, manufacturing systems analysis, and machine maintenance. Prior to joining IBM, he was a Senior Member of Research Staff at Philips Laboratories, the North American research arm of Royal Philips Electronics of the Netherlands. Ying Tat holds a B.Sc. degree from the University of Hong Kong, M.S. and Ph.D. degrees in Industrial Engineering from the University of Wisconsin - Madison.

Grace Lin *IBM Business Consulting Services, P.O. Box 218, Yorktown Heights, NY 10598 (gracelin@us.ibm.com).* Dr. Grace Lin is an Executive Consultant and an Associate Partner in IBM's Business Consulting Services. She is also the Global Sense and Respond leader and an elected member of the IBM Academy of Technology. Prior to joining IBM Consulting in Feb, 2003, she served as the Senior Manager of Supply Chain management and the e-Business Optimization Department and the Program Director of the Value Chain Innovation Center at the IBM T. J. Watson Research Center. Dr. Lin led her team to win the INFORMS Franz Edelman Award in 1999 and was a recipient of the Purdue Distinguished Industrial Engineer Award in 2003. She was elected as INFORMS VP Practice twice and serves on the INFORMS Board of Directors. Dr. Lin is also on the Editorial Board of OR and MSOM.

Robin Lougee-Heimer *IBM Research Division, Thomas J. Watson Research Center, P. O. Box 218, Yorktown Heights, New York 10598 (robinlh@us.ibm.com).* Dr. Lougee-Heimer is a Research Staff Member in Optimization and Mathematical Software at the IBM Thomas J. Watson Research Center, in Yorktown Heights, NY. She joined IBM Research in 1994 after completing her PhD in Mathematical Sciences from Clemson University. Dr. Lougee-Heimer's research activities and interests include applications of optimization to solve large-scale business problems.

Rakesh Mohan *IBM Research Division, Thomas J. Watson Research Center, P.O. Box 218, Yorktown Heights, NY 10598 (rakeshm@us.ibm.com).* Dr. Mohan is the Senior Manager for Industry Solutions at IBM T.J. Watson Research Center and is involved in building innovative solutions for retail, automotive and other industries. Previously, he managed the E-Commerce Platforms group and was involved in building e-commerce, e-marketplace and procurement systems. Rakesh Mohan joined IBM Research in 1989 after receiving a PhD and MS in Computer Science from the University of Southern California.

Chris Nokkentved *IBM Business Consulting Services, Nymøllevej 91/4, 2880 Lyngby, Denmark (chris.nokkentved@dk.ibm.com).* As an Executive Consultant in IBM's Business Consulting Services, Mr. Nøkkentved is currently responsible for the SAP Alliance in EMEA, as well as the Supplier Relationship Mgmt (SRM) solution area in EMEA. He has also been part of the team that established the SAP SCM practice in EMEA, and has spend 3+ years researching and leading projects in the area of collaboration (B2B eMarkets, eSCM and SRM). Prior to that he had 9+ years of experience with ERP systems. Mr. Nøkkentved received a M.Sc. degree (with honors) in Business Administration and Computer Science from the Copenhagen Business School, an MBA degree from the University of Texas in Austin, and is currently finishing his Ph.D. studies on "Enabling Value-Creation with ICT in Collaborative Supply Networks". He has been involved in developing various methodologies for Collaborative eMarkets (via CPFR, SCOR, and Integration eHubs - RosettaNet), lectured in various universities (CBS, MIT), and is frequently presenting in research and business conferences (i.e. Strategic Management Society, International Marketing & Purchasing Group). He has published research articles and a book in the field of MNCs and Management Processes.

Jo van Nunen *Rotterdam School of Management, Erasmus University Rotterdam, P.O. Box 1738, 3000 DR, Rotterdam (J.Nunen@fbk.eur.nl).* Dr. Van Nunen is the chairman of the Department of Decision and Information Sciences of the Rotterdam School of Management. He is also program leader of the research program on logistics and information systems, which is a joint program with the Technical University Delft and Erasmus University Rotterdam. He is the scientific director of a national research program on "Transition to Sustainable Mobility". Currently his research focuses on closed-loop supply chains and on ICT applications in logistics. Many of the research projects he is involved in are co-operations with private companies and governmental organizations. As can be expected, logistic organizations in the Port of Rotterdam are an important source of inspiration for his research. Dr. Van Nunen is also a part time advisor for Deloitte.

Heiko Pieper *Vivecon Corporation, 650 Castro Street, Suite 300, Mountain View, CA 94041 (heiko.pieper@vivecon.com).* Heiko Pieper is one of the early members of the Vivecon Corporation, a pioneer in delivering Supply Chain Risk Management solutions. He helped develop the analytical methods underlying the Vivecon technology and has extensive experience with Vivecon's clients in the Automotive Industry in Europe and the US. Prior to joining Vivecon, Dr. Pieper received his PhD from Stanford University (Management Science & Engineering) where he was part of the Systems Optimization Laboratory. During this period, his research focused on methods to efficiently solve multi-level and stochastic optimization problems and their application to deregulated electricity markets.

David J. Seybold *IBM Business Consulting Services, 100 East Pratt Street, Baltimore, MD 21202 (dseybold@us.ibm.com)*. Partner and Vice President, Supply Chain Management. David is presently responsible for supply chain consulting and services in the Distribution Sector industries including Consumer Goods, Pharmaceutical, Retail, Travel related industries. David has over 15 years experience in supply chain and product development consulting and solutions across a variety of industries. He received a B.S. in Quantitative Analysis and a B.A. in Economics from Penn State University and an M.S. in Operations Research and an MBA from the University of Maryland Smith Business School.

Dailun Shi *Department of Management Science, School of Management, Fudan University, 670 Guoshun Road, Shanghai 200433, P.R. China (dlshi@fudan.edu.cn)*. Dr. Shi is a Professor and Research Fellow in the School of Management at Fudan University. He received his Ph.D. in Management, MSM (MBA) in Finance, and M.S. in Industrial Engineering, all from Georgia Institute of Technology. He also earned an M.S. degree in Mathematics from Brown University. Prior to his career at Fudan, Dr. Shi worked as a Research Staff Member at IBM Thomas J. Watson Research Center in New York, and as a manager at Citigroup headquarters for network crisis management. His current research areas include Supply Chain Management and Logistics, Risk Management, Operations Management, and e-business. He has applied for 14 patents in the US, and published papers in such journals as Management Science, IBM Systems Journal, IIE Transactions, Supply Chain Management Review, and Journal of Systems Science and Systems Engineering.

Moninder Singh *IBM Research Division, Thomas J. Watson Research Center, P. O. Box 218, Yorktown Heights, NY 10598 (moninder@ us.ibm.com)*. Mr. Singh is a Research Staff Member at the IBM T.J. Watson Research Center. He joined IBM in 1998 after receiving a Ph.D. in Computer and Information Science from the University of Pennsylvania. His research interests include machine learning and data mining, especially their application to problems in sourcing and procurement, knowledge representation, probabilistic reasoning, and privacy technologies. Moninder received a B.Tech. in Computer Science and Engineering from the Indian Institute of Technology, Delhi in 1991, and an MS in Computer Science from the University of South Carolina in 1993.

Stavros Stefanis *IBM Business Consulting Services, 136 Crawley Falls Rd, Brentwood, NH 03833 (stefan1@us.ibm.com)*. Dr. Stefanis is a Partner in IBM's Industrial Embedded System Lifecycle Management practice.

Stavros is responsible for helping clients transform their engineering organizations to gain time to market and productivity benefits. He has experience in the automotive, aerospace and electronics industry and manages the engineering design/embedded development skills for a group of 60 practitioners in IBM. Stavros has nearly 13 years of software development and systems engineering consulting experience including significant time dedicated to process modeling and systems decomposition research and development. Stavros has published over 20 full length papers in many business and academic magazines. Stavros has also published a chapter entitled "Product Lifecycle Management" on Webster's encyclopedia of electrical engineering. Dr. Stefanis has been with IBM for 5 years where as a supply chain competency leader he has developed skills in product strategy & architecture, conceptual design, systems engineering, forecasting, parts management, planning and collaboration. He is also the executive supply chain advisor for many large automotive OEMs. Before joining IBM, Stavros has architected and implemented a variety of engineering, supply chain and industrial control automation products. Stavros has co-developed the industry standards for artifact collaboration as a leading member of the Rosetta-Net committee and is a member of AIAG for the development of engineering standards for systems development. He has also been an advisor to IBM's Rational Software competency in the areas of system decomposition and integrated requirements management.

Alex E.D. Suanet *IBM Netherlands, Johan Huizingalaan 765, Amsterdam 1066, The Netherlands (alex.suanet@nl.ibm.com).* Mr. Suanet is an advisory IT Specialist in the area of Business Analysis at IBM Application Maintenance Services in Amsterdam, The Netherlands. He is working in external projects as Business Information Analyst and he is advising customers on improving supply chain processes. He started his career in 1997 at IBM within EMEA Service Logistics (ESL) in Amsterdam. The business unit ESL is responsible for the spare parts logistics in EMEA. He designed in a team the Forecast and Planning modules for a Global part system. This system controls the spare logistics process of IBM globally. In 1997 he finished his master in Econometrics with as specialty Operations Research at Tilburg University.

Ko-Yang *Wang IBM Global Services, 294 Route 100 Somers, NY 10589 (kyw@us.ibm.com).* Dr. Ko-Yang Wang is an IBM Distinguished Engineer and a Research & Innovation Executive in Business Transformation, IBM Global Services. He also leads a cross functional team of more than 45 IBM fellows and distinguished engineers on "Enterprise of the Future" which explores 2-5 year visions of business/technology game changers to help IBM and its clients set strategies. His current interest is on the Fusion and Business and IT and its impacts on business. Prior to his current position, he was a Principal in the IBM Consulting Group, the chief architect and senior manager for IBM Global Services' Knowledge Management technologies. He was a team leader and a Research Staff Member in IBM Research from 1991-1996. He led his team to win the 1998 and 1999 Giga Excellence Award on KM and Workflow. He was also awarded an Outstanding Technical Achievement award and an IBM Consulting Group Division award. Dr. Wang received his Ph.D. and Master degrees in Computer Sciences from Purdue University. He also has a M.S. and a B.S. degree in Mathematics.

Robert Wittrock *IBM Research Division, Thomas J. Watson Research Center, P.O. Box 218, Yorktown Heights, NY 10598 (wittrock@ us.ibm.com).* Dr. Wittrock is a research staff member in the Mathematical Sciences department at IBM Research. He is the primary designer and implementer of the implosion software known as WIT. His previous work has been in the area of applying optimization and heuristic techniques to various supply-chain management problems and in the area of large-scale optimization. He received his PhD degree in operations research from Stanford University in 1983.

David D. Yao *IEOR Dept, Columbia University, New York, NY 10027 (yao@ieor.columbia.edu)* received his Ph.D. degree from the University of Toronto in 1983, and started his academic career at Columbia University, where he became a full professor in 1988. He is an IEEE Fellow, the Stochastic Models Area Editor of Operations Research, and Editor-in-Chief of Foundations and Trends in Stochastic Systems. He has done extensive research and consulting work in stochastic networks, semiconductor manufacturing, and supply chain management.

14 Index